Advances
in Biochemical
Engineering 1

Edited by

T. K. Ghose and A. Fiechter

With 70 Figures

Springer-Verlag
Berlin Heidelberg GmbH 1971

T. K. Ghose

Dept. of Chemical Engin., Indian Institute of Technology, New Delhi/India

A. Fiechter

Mikrobiologisches Institut der Eidgen. Techn. Hochschule, Zürich/Schweiz

ISBN 978-3-662-15596-7   ISBN 978-3-540-36528-0 (eBook)
DOI 10.1007/978-3-540-36528-0

Advances in Biochemical Engineering 1

Edited by

T. K. Ghose and A. Fiechter

With 70 Figures

Springer-Verlag
Berlin Heidelberg GmbH 1971

T. K. Ghose

Dept. of Chemical Engin., Indian Institute of Technology, New Delhi/India

A. Fiechter

Mikrobiologisches Institut der Eidgen. Techn. Hochschule, Zürich/Schweiz

ISBN 978-3-662-15596-7   ISBN 978-3-540-36528-0 (eBook)
DOI 10.1007/978-3-540-36528-0

Foreword

Biochemical engineering is concerned with the effective practical application of the knowledge amassed by the biological sciences. The effort to make the industrial operation of a biological process effective and economically viable involves substantial interaction between chemical engineers, biochemists, microbiologists, and geneticists. This chemical engineering hybrid has given rise to new and distinctive unit operations and systems for bioprocesses: sterilization, oxygen transfer, continuous culture, cell separation, biofiltration, transport of biological fluids, biopolymerization, enzyme insolubilization, novel separation techniques etc. We also have a better understanding of catalysis, kinetic models and scale-up steps involved in microbiological processes. Over the past few decades, the biological scientists have produced vast amounts of quantitative information. The life sciences are now seeking a unified basis, with exact knowledge replacing the descriptive approach. Many biological phenomena are now understood and can be employed for the benefit of mankind.

The outputs of biotechnology represent wealth generated by the biochemical and fermentation upgrading of the cheaper kinds of natural resources. Surplus agricultural and industrial by-products constitute these resources. While in many cases it has been possible to achieve spectacular reductions in microbiological production costs, the risk involved in starting a microbiological venture has never been small, primarily due to lack of knowledge and talents. Once the problem is recognised for what it is, we may see that a realistic solution lies in breaking down barriers to communication. This will attract new talents to contribute in biochemical engineering research and thereby help advance biotechnology.

This book is the first in a projected series of multi-author books concerned with the engineering aspects of biotechnology. It is intended to serve the established professional and also to encourage students to take up careers in this field. The mathematical analysis of microbiological systems and bioprocesses will be given prominence in this series. The first volume deals with some hitherto neglected areas: the dynamics of fermentation fluids, enzyme applications, internal regulation of metabolic systems, novel energy and carbon sources, advances in the engineering analysis of cell separation systems, and the kinetics of the break-down of solid substrates. The editors hope that these contributions will stimulate many more talents to contribute through basic research and dissemination of knowledge to the "yet to emerge" hybrid discipline of biochemical engineering.

October, 1970
New Delhi, India
Madison, Wisc., USA

T. K. Ghose
A. Fiechter

Contents

CHAPTER 1

The Nature of Fermentation Fluids

HISAHARU TAGUCHI

With 15 Figures

Contents

Introduction

Microrganisms in submerged culture fermentation thrive in a liquid environment which supplies them with the nutrients they need for biochemical activity. This liquid not only serves as a reservoir for substrate, dissolved oxygen and metabolic products, but also absorbs the heat liberated by the reacting cell. The rate of transfer of heat and materials to and from the cells depends largely on the properties and motion of the fermentation fluids in which the transport is carried out. Therefore, a knowledge of the properties of fermentation fluids is an important prerequisite for an intelligent understanding of these transport processes and for analyzing the methods of process design, scale-up and continuous processing. An understanding of the properties of fermentation fluids is required for the study of the following important problems in fermentation technology:

1. Development of equipment to allow more efficient gas absorption in gas-liquid dispersion systems.

2. Economic fermentation processes of high microbial solids.

3. Evaluation of shear in the design of stirred fermenters.

4. Transport and determination of residence-time distribution function in viscous continuous fermentation.

5. Growth in cultures with two liquid phases.

Although information about the physical nature of fermentation fluids e.g., cell size, hindered settling of microbes, viscosity of fermentation broths and interfacial tension of fermentation liquids ... etc., is indispensable for the design and operation of all types of fermentation processes, there were relatively few data upon which calculations for design and scale-up could be based. At the present time, the biochemical engineer is making important contributions to the design and scale-up of fermenters by investigating the physical properties of fermentation fluids and accumulating these data, as well as by obtaining a basic understanding of fermentation kinetics.

However, there is a wide variety of complicated fermentation systems. Some of them are quite insensitive to the detailed nature of fermentation fluids, and at times a similar concept used for chemical processes can be applied to the design of fermentation processes. For example, aerobic fermentations involve mixing, mass transfer and heat transfer which are unit operations familiar in chemical engineering. The particular case of fermentation, in which living bacteria and fungi are affected by fluid shear, and in which high mass-transfer rates require large expenditures on equipment and power, has all the features necessary to warrant a thorough study of mixing mechanisms based on the nature of fermentation fluids. Many physical problems are also found in hydrocarbon fermentation. The objective of this paper is to present an up-to-date review of the nature of fermentation fluids and an appraisal of the use of these physical properties in engineering and analytical correlation employed in fermentation practice. However, the problems dealt with are limited and rather particular. The application referred to in this paper should be of help in furthering studies of biochemical engineering problems associated with the nature of fermentation fluids.

1. Rheological Properties of Fermentation Fluids

In many cases, a fermentation process must be a high solids process in order to be economic. Substrate, microbes, or product, or all three together may be considered as solids. In general, fermentation broths of mold and streptomyces exhibit non-Newtonian behavior, either Bingham or pseudoplastic, so that the rate of heat and mass transfer in the fermentation process depends largely on the rheological properties of the fermentation fluids.

Deindoerfer (1960) and Richards (1961) have published excellent reviews on this subject, and presented an appraisal of the use of these properties in engineering and analytical correlations employed in fermentation practice. Although there are some cases where the operation and scale-up of these mycelial fermentations have been accomplished successfully using either the power input per unit volume or the overall oxygen transfer coefficient of Newtonian fluids as a basis, this does not preclude further study of the scale-up of non-Newtonian fermentation broths. On the contrary, much more information about properties such as oxygen transfer in bubble aeration, and mixing time in non-Newtonian fermentation fluids is needed to provide a better understanding of the operation and scale-up of fermentation processes.

a) Controlling Operation by Fluid Viscosity and Temperature

Carilli, Chain, Gualandi, and Morisi (1961) have reported that the aeration rate obtained with pellets was more than twice that obtained with filamentous mycelium in the same amount of mycelial fluid. This effect is due mainly to the rheological properties of the mycelial fluid which are shown to vary within wide limits according to the morphological form of the mycelium. For example, observations range from about 500 pseudocentipoise with the pellet form to over 5000 pseudocentipoise when the mycelium is filamentous. Correspondingly, the flow behavior index value is 0.4 with pellet growth, and decreases to almost zero when the mycelial growth is filamentous, showing that under the latter conditions the mycelial fluid exhibits entirely non-Newtonian behavior. Moreover, the rheological properties of most fermentation fluids change during the course of fermentation, e.g. Taguchi and Miyamoto (1966) have found that the value of the flow behavior index varied from 0.3 to 0.9 and the consistency index varied from 2 to 35 dyn/cm^2 secs with a glucamylase fermentation by *Endomyces* species. When a very viscous fluid of filamentous mycelium is diluted by as little as 10 or 15% with water or feeding medium, a steep drop in the consistency index to about half the previous value and a corresponding marked increase of the oxygen transfer rate are observed, and this is reflected by a sudden increase in the concentration of dissolved oxygen.

Satoh (1961) has studied the rheological properties of kanamycin fermentation broth which displayed typical Bingham behaviour in the course of fermentation, so that it became difficult to disperse the nutrients uniformly and to supply oxygen to the microbial population.

1*

When a kanamycin broth was diluted with 5% by volume of sterile water to reduce the viscosity, the yield increased by 20%. Therefore, it may be possible to improve the performance of a fermentation by controlling its viscosity on the basis of an understanding of the rheological properties of the fermentation fluid.

An investigation was made of the influence of temperature on the rheological properties of kanamycin fermentation by Satoh (1963), and the method of reducing viscoelasticity was applied to a non-Newtonian fluid. Andrade (1930) obtained already the following relation between the viscosity and temperature using the energy barrier theory.

$$\eta = A\,e^{E/RT} \tag{1}$$

where E is the activation energy, T is absolute temperature, R is a gas constant for the material and η is the viscosity. Then the following equation can be introduced.

$$E = \frac{R\,d(\ln\eta')}{d(1/T)} = 2.303\,R\,\frac{d(\log\eta')}{d(1/T)} . \tag{2}$$

First of all oleum zinc oxide, as a typical Binghamplastic example, was rheologically analyzed. Its plastic viscosity was reduced with increasing temperature, but no change in yield shear stress was observed. When $\log\eta'$ was plotted against $1/T$ the relationship appeared quite linear, where η' is the plastic viscosity. The same relationship was obtained with olive oil which was the dispersion medium. Therefore, it was concluded that the activation energy of oleum zinc oxide at any temperature is equal to that of olive oil. Then, it was examined whether the above-mentioned relationship between temperature and plastic viscosity or yield shear stress would apply for kanamycin fermentation broths. The plastic viscosities throughout the kanamycin fermentation periods, within the same temperature range as the case of oleum zinc oxide, were examined, and it was found that these were consistent with the above-mentioned observations throughout the whole kanamycin fermentation period. The gradients of the straight lines, $\log\eta'$ against $1/T$ observed with broths of various ages of fermentation were found to be equal as shown in Fig. 1. Therefore, plastic viscosity at any time may be calculated. Next the effects of temperature on kanamycin production were studied. The time of change of temperature was programmed on the basis of the agitated culture experiments (the optimum growth temperature is 30° C and the optimum kanamycin production tempera-ture is 20° C): (1) mycelium was grown at 28° C for 40 hours after inoculum and changed to 24° C, (2) the control run was incubated at 28° C for entire fermentation. It is evident from Fig. 2 that in the fermentation at lower temperature the rate of consumption of sugar

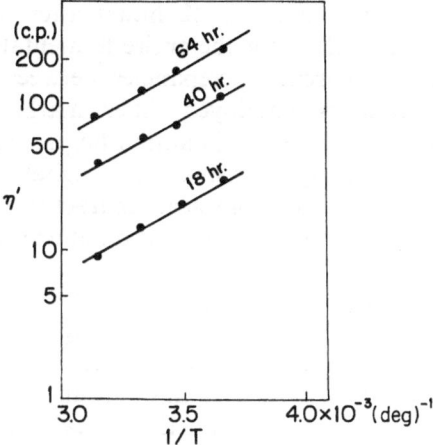

Fig. 1. Logarithmic plot of plastic viscosity against $1/T$

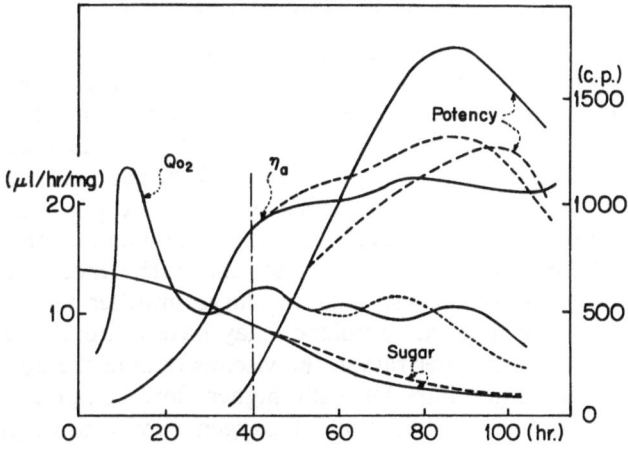

Fig. 2. Time course of kanamycin fermentation

was somewhat slower than with fermentation conducted at the higher temperatures, but the apparent viscosities of the broths showed contrary results. And it was shown that the production of kanamycin was 35% less in a fermentation run which was started at 28° C and changed to 24° C at 40 hours after inoculation, than, when incubated at 28° C during the whole fermentation course.

On the other hand, Owen and Johnson (1955) have reported on the effect of temperature changes on the production of penicillin, and it was found that 50% more penicillin was produced by fermentations started at 30° C and changed to 20° C 42 hours after inoculation, than in controls incubated at 25° C for the entire fermentation period.

Here we have apparent discrepancies between the results of two studies. The organisms employed differ markedly in oxygen and nutrient requirements as well as in antibiotic production. Furthermore, the morphological changes arising from agitation result in broths markedly different in their physical characteristics, and it is also extremely difficult to compare the aeration efficiencies at each stage of two fermentations.

However, it is clear that attempts to correlate antibiotic production with the effect of temperature changes during fermentation should be carried out with an understanding of the correlation between the rheological properties of fermentation fluids and mass transfer, nutrient requirements, and the rate of production at each stage of fermentation.

b) Effects of Geometric Parameters in Viscous Fermentation Fluids

α) Ratio of Impeller Diameter to Fermenter Diameter

Sulfite oxidation rate or power input are not always proportional to the simple oxygen transfer rate actually available in non-Newtonian fermentation fluids. Several workers have shown how added mycelium or other fibrous materials reduce the oxygen transfer measured in non-viscous fluids. The ratio of impeller size to fermenter size in viscous fermentation broths has a pronounced effect on oxygen transfer and blending. Namely, the flow component of the power consumption (proportional to ND_i^3) is greater and the turbulence component (proportional to $N^2D_i^2$) is less with the large impeller than with the small impeller. The reduced turbulence may have a more pronounced effect on the oxygen transfer rate in the viscous than in the non-viscous system, and the larger impeller with higher flow component would effect a more uniform distribution of oxygen supply throughout the fermenter. The smaller impeller causes local high turbulence and increased oxygen transfer in viscous fermentation fluids. These situations were demonstrated clearly by Steel and Maxon (1962) in studies of novobiocin fermentation which exhibits plastic behavior. At constant agitation speed the oxygen availability rate (OAR) decreased with increase in apparent viscosity of novobiocin fermentation broths, but the magnitude of the reduction was less with a small impeller than with a large one. Optimum novobiocin yields were achieved at a power input of 0.5 HP/100 gal. when the impeller diameters were 29 or 39% of the fermenter diameter. A power input of 0.75 HP/100 gal. was required for equivalent results with the impeller diameter 49% of the fermenter diameter. At constant power input the smallest impeller gave an OAR about eight times that of the largest one in viscous fluids as shown in Fig. 3, whereas with sulfite or non-viscous fluids all impeller diameters gave the same OAR. They observed that OAR measured in actual novobiocin fermentation can be used to correlate antibiotic yields, and that impeller tip velocity can be used to correlate with OAR, independently of impeller diameter, in a viscous fermentation system. For the illustration of these results, they described an analogous correlation between OAR and impeller tip velocity in other systems.

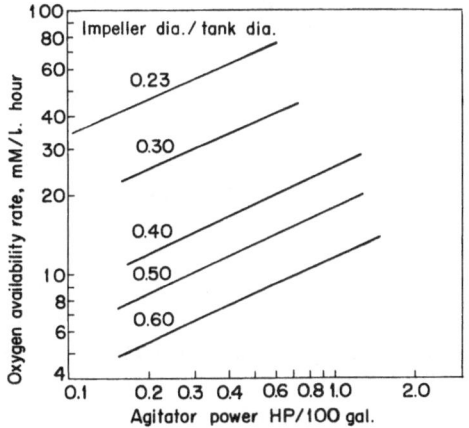

Fig. 3. Effect of power on oxygen availability rate in novobiocin fermentation

For example, Fondy and Bates (1961) found that, in a stirred tank, the particle size of a sodium-potassium alloy in mineral oil varied with tip velocity and that it was raised to the -1.8 power. In Steel and Maxon's work, OAR varied with tip velocity raised to the 1.6 power. This is in close agreement and it indicates, as might be expected, a close inverse relationship between oxygen transfer rate and bubble size. Finn has noted a correlation between impeller tip velocity and oil droplet size in stirred suspensions which were stabilized with surface-active agent. In both systems of novobiocin fermentation and oil suspensions with added surface-active agent, it is apparent that the particles or bubbles do not recoalesce; on the other hand, without added surface-active agent in the oil suspensions the oil droplets recoalesce, and droplet size no longer correlates with tip velocity but with power input instead. The similarity already noted between these systems and aerated non-Newtonian fermentation fluids suggests also that the air bubbles in these fermentation broths do not readily coalesce. The fact that sulfite oxidation values correlate with power input and not with impeller tip velocity may be due to a relatively rapid recoalescence of air bubbles in sulfite solutions.

To obtain a high rate of oxygen transfer the interfacial area between gas and liquid must be large, therefore, the bubble size must be small. In a coalescing system, sulfite solution, it is necessary to supply sufficient power to maintain small bubbles. In a noncoalescing system, viscous fermentation, it is only necessary to provide a zone of high shear to assure the initial formation of small size bubbles. Therefore on the basis of OAR per unit power input, the optimal impeller size for viscous fermentation is probably the smallest size that would give an adequate circulation rate of the fermentation broth without causing physical disruption of the organism.

Another reason proposed for the greater efficiency of small high-speed impellers was that they were better able to break up mycelial clumps and hence reduce the magnitude of the oxygen transfer resistance. The true critical dissolved oxygen concentration for most cultures is between zero and 10% of saturation with air at atmospheric pressure. For viscous fermentation broths, Phillips and Johnson (1961) found that a high apparent critical dissolved oxygen concentration occurred in one part of the fermenter, while the respiration rate was limited by low dissolved oxygen concentration in other parts of the fermenter. They attributed this to a lack of good bulk mixing, and to the resistance to oxygen transfer offered by mycelial clumps. The studies of Steel and Maxon (1966) confirmed the work of Phillips and Johnson inasmuch as they also reported that mixing was non-homogeneous with small diameter impellers in novobiocin fermentation. Bulk mixing was found to be incomplete in 20-liter pilot fermenters with impellers of $D_i/D_t = 0.4$, but appeared homogeneous with impellers of $D_i/D_t = 0.69$. In the small-impeller system, the respiration rate was presumably limited by insufficient oxygen supply in parts of poor bulk mixing, whereas in the large-impeller system the major resistance was between the bulk of the liquid and the cell, i.e. intraclump resistance. Thus, the apparent critical dissolved oxygen concentration indicated by the dashed line in Fig. 4 differed according to the ratio of impeller diameter to tank diameter. With large-diameter impellers bulk mixing was much improved, but the apparent critical dissolved oxygen concentration was high, 50 to 60% of saturation, whereas the true critical level for this culture is below 10% of saturation. From these results it was inferred that a major resistance to oxygen transfer existed within the mycelial clumps. Phillips and Johnson estimated that the clump diameter must be of the order, of 0.83 mm in order to show a significant effect of limiting the oxygen transfer. These finding are consistent with flow behavior as impeller diameter is changed. Large impellers exhibit high flow or pumping capacity but a low shear component; hence, because they are less able to break up mycelial clumps, the relative contribution of clumps to oxygen transfer resistance becomes greater. In production-scale fermenters, measurements of dissolved oxygen indicated that bulk mixing was complete with each of three impeller sizes tested ($Di/Dt = 0.28$, 0.33, and 0.43), but that the respiration rate was again limited. Hence, it behaved in about the same way as a pilot fermenter with large impellers, and the apparent critical dissolved oxygen concentration was high. Here, too, oxygen transfer was limited mainly by resistance between the bulk of the liquid and the cell.

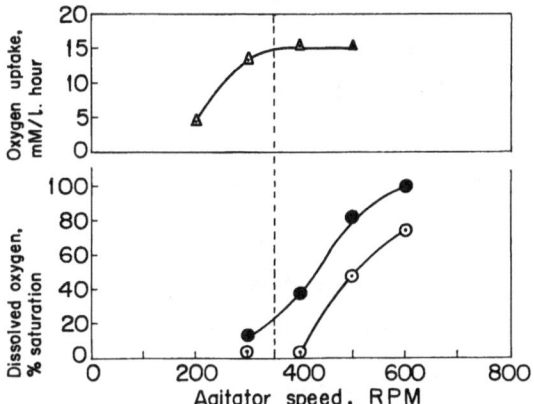

Fig. 4. Effect of agitator speed on oxygen uptake and dissolved oxygen in 20 l novobiocin fermentation with impellers of $D_i/D_t = 0.4$, apparent viscosity 300 cp; air flow rate 5 slm; lower probe (●) opposite the lower impeller, behind a baffle; upper probe (○) near liquid surface behind a baffle

Steel and Maxon (1966) have attemped to solve the problem of supplying adequate oxygen through studies of impeller design, and found the advantages of the multiple-rod impeller shown in Fig. 5. It gave the same novobiocin yield and oxygen availability rate at about one-half of the power required by turbine impeller.

Blakebrough and Sambamurthy (1966) carried out oxygen-transfer studies in gas-liquid and gas-liquid-solid systems, and found significant differences in the operation of turbine impellers in the two systems. In the former system, at a given aeration rate, a limitation in the operation of the impellers was reached at a power input of 0.36 HP/m³. Both below and above this level, the efficiency of oxygen transfer was independent of the manner in which power was transmitted, namely, irrespective of impeller speed, impeller diameter, blade dimensions, and blade configuration. Below this level, changes in oxygen transfer rate were mainly associated with changes in bubble size, while beyond this, retention time was the more significant factor. However, with the solid filamentous phase, the pattern of the results changed significantly. The remarkable result from the tests on the three-phase system is that, in so far as oxygen transfer is concerned, at any given power level it is

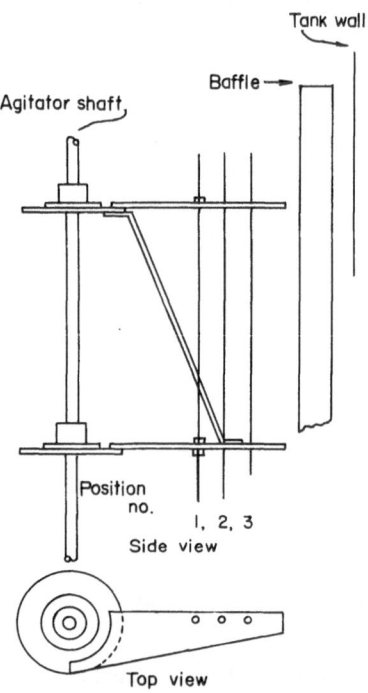

Fig. 5. Multiple-rod impeller — production design

advantageous to use impellers with small blades and moderate overall diameter at moderate rotational speeds, or to use small blades in combination with large overall diameter operating at relatively high rotational speeds, or to use small blades in combination with large overall diameter and moderate rotational speed. The result is applicable up to a power input of about 0.9 HP/m^3.

In both systems, mixing times could be related to the momentum factor, $ND_i \times NWL(D_i - W)$, over a wide range of variables, where W is impeller blade width and L is impeller blade height. The correlation obtained in the gassed system under nonflooding conditions was parallel to that obtained in the non-gassed system. With the onset of flooding, mixing times rapidly increased, presumably due to the force set up by the escaping air bubbles, which oppose those produced by the rotating impeller. The data obtained in the three-phase system can be grouped in the form of the following empirical equations.

$$K_L a = 152 \left[t_m^{0.203} (P_g/V)^{1.79} / M_f^{1.05} N^{0.046} \right] \tag{3}$$

$$K_L a / N = 0.113 \left[\frac{D_t^2 H_L}{WL(D_i - W)} \right]^{1.43} \left(\frac{1}{Nt_m} \right)^{1.09} \left(\frac{D_i}{D_t} \right)^{1.02} \tag{4}$$

where P_g/V is power input per unit volume, $K_L a$ is overall mass transfer coefficient, M_f is momentum factor, t_m is mixing time, H_L is height of liquid and D_t is tank diameter. Mixing time, which is considered an important criterion in fermentation processes, has been included in these equations. The overall mass transfer coefficient has been related to impeller geometry and operating variables. In the first equation, absolute values, rather than dimensionless groups, are employed, and this equation can be used for varying oxygen transfer rates in a fermenter of a particular size. The second equation is likely to be a more useful form for studies on the scale-up of fermenters.

β) Effect of Impeller Spacing

The available literature contains very few references to studies of the power consumption for agitation and the volumetric oxygen transfer coefficients of multi-stage impeller fermenters. Takeda and Hoshino (1968) showed experimentally that closely spaced impellers caused serious interference between the flow streams from adjacent impellers and an overall reduction in power consumption. On the other hand, Oldshue (1966) has indicated that within fairly large ranges of geometric variables, a similar oxygen transfer coefficient is obtained if the power per unit volume is maintained at similar values under a constant aeration rate.

Richards (1963) has reported qualitatively the effect of varying impeller spacing on flow patterns and power consumption in a fermenter as follows.

In fully baffled tanks, the types of turbine impeller result normally in a flow of liquid from the central axis of the tank to the walls, then vertically up or down, being finally drawn back again into the eye of the impeller. If the impellers are too widely spaced, it becomes possible, especially in such liquids as viscous fermentation broths, to have regions unagitated by the action of the impellers. On the other hand if they are too close together, serious interference can occur between the flow streams from two adjacent impellers. This can have the dual effect of reducing power input and inadequate mixing.

Fig. 6, presented by Richards, shows the effect of varying impeller spacing on flow patterns in a pilot-scale fermenter broth. These patterns are indicative of what can occur, even in large tanks. In Fig. 6 (a) the impellers are too close together and clearly act as one wide impeller. Flow to the middle impeller is restricted, power input is reduced, and the tank is inadequately agitated. In Fig. 6 (b) a small increase in impeller spacing results in no further increase of power input, but some

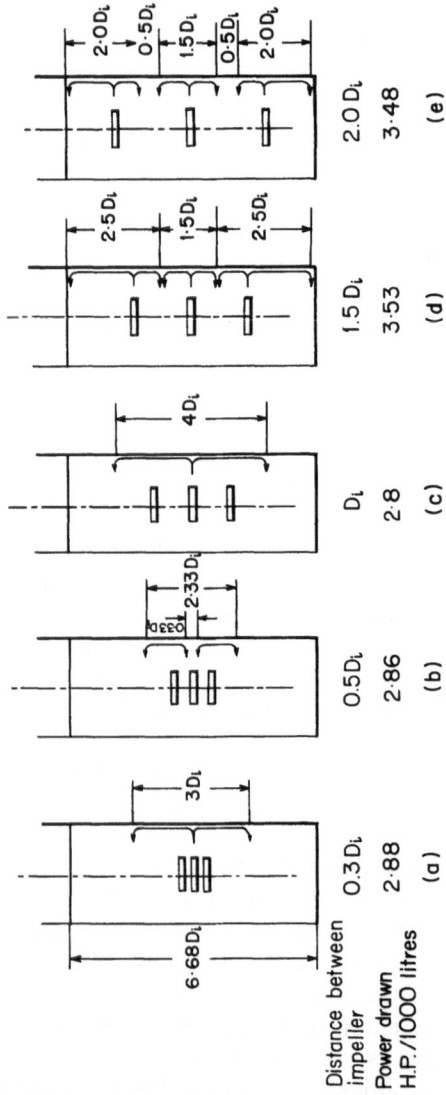

Fig. 6. Effect of impeller spacing on flow pattern and power absorption

independent flow to the middle impeller causes the discharge from it to interfere with the flow from the other impellers, giving an overall reduction in the height of liquid agitated. Fig. 6 (c) shows the impellers spaced one impeller diameter apart. The height of liquid agitated again increases but the power input remains substantially unchanged. At this spacing in water or in non-viscous Newtonian fluid the impellers would usually be working nearly independently with the return flows touching and mixing as in Fig. 6 (d). However, in viscous non-Newtonian fermentation broth a pattern as shown in Fig. 6 (c) can occur. When the impellers are 1.5 diameters apart, as in Fig. 6 (d), good mixing of the whole tank contents is obtained. No stagnant areas are apparent between impellers and a significantly greater amount of power input is observed. Fig. 6 (e) shows the effect of spacing impellers too widely apart. In fermentation broths, the rheological characteristics of which are less strongly non-Newtonian, impellers can be operated satisfactorily at more than two diameters apart. In the case illustrated in Fig. 6 (e) the limit is being approached for the broth in question and stagnant areas are beginning to appear. The power has reached a maximum and is unchanged by further separation of impellers. The correct spacing of impellers is affected largely by the rheological characteristics of the fermentation broth being agitated. Of course, at any given impeller spacing a transition from one pattern to another can occur as the fluid characteristics change during the course of a fermentation. In borderline cases, there appear to be critical impeller spacings, which can result in unstable flow patterns and the power input may fluctuate as the pattern changes. Changes of this type can sometimes be relatively sudden. Taguchi and Kimura (1970) studied the effects of varying impeller spacing on power consumption and volumetric oxygen transfer coefficient in a 100-liter stirred tank, both with and without aeration. In the non-gassed system, the maximum power consumption of an i-stage impeller system was observed with impeller spacing determined by the following equation;

$$S_i = \frac{\log i}{2\{(i-1)\log i - \log(i-1)!\}} \cdot \left[H_L - \left\{ 0.9 + \frac{(i-1)\log i - \log(i-1)!}{\log i} \right\} D_i \right] \tag{5}$$

where S_i is impeller spacing, H_L is liquid depth, D_i is impeller diameter, and i is the number of impeller stages.

In non-viscous Newtonian fluids, the correlation between gassed power consumption and non-gassed power consumption observed by Michel and Miller (1962) was found to be applicable to multi-stage impeller systems. The maximum volumetric oxygen transfer coefficient was obtained under the impeller spacing which results in the maximum power consumption in non-gassed system at any given impeller diameter, rotation speed, and aeration rate.

The effects of increasing the number of impeller stages on the volumetric oxygen transfer coefficient were also studied, using the optimum impeller spacings calculated from Eq. (5). In this experiment, deep liquid depth was employed, i.e. the ratio of liquid depth to

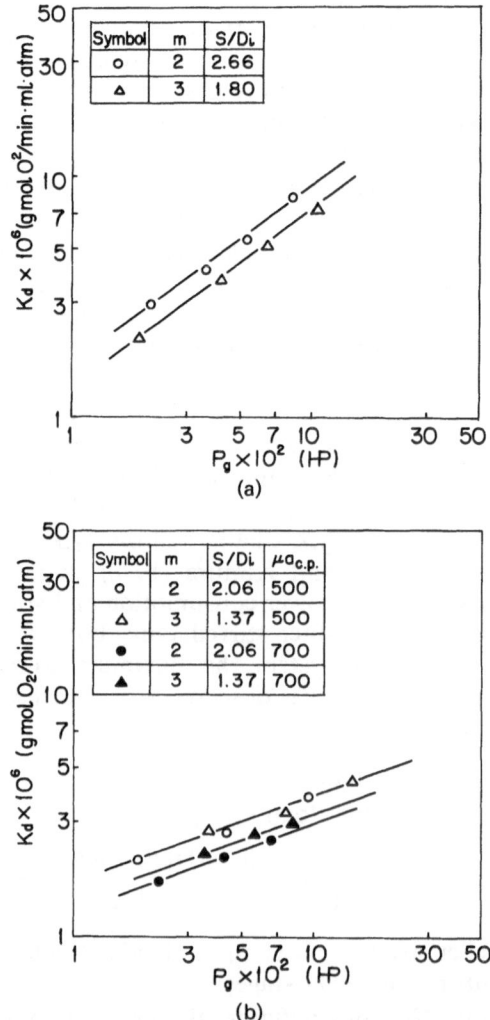

Fig. 7. The effects of increasing the number of impeller stages on the volumetric oxygen transfer coefficient. (a) in water system, (b) in viscous mycelial suspension

fermenter diameter was 2.41. When sulfite solution was used, the volumetric oxygen transfer coefficient was considerably higher in the two-stage system than in three-stage system, as shown in Fig. 7a. A positive effect, however, of increasing the number of impeller stages on the volumetric oxygen transfer coefficient was observed with a mycelial suspension of high apparent viscosity, about 700 centipoise, as shown in Fig. 7b. Thus the correct spacing of impellers and the number of impeller stages are matters which depend largely on the rheology of the fermentation broth being agitated. Therefore, further studies of the correct impeller spacing in viscous fermentation broths must be made from this point of view.

c) Power Consumption in Gassed Systems

The problem of predicting the power consumption of agitators operating in non-Newtonian fluids has been reviewed by Metzner (1960), and reasonably certain methods exist. The problem only arises when very viscous fluids are being mixed, since otherwise in fully baffled tanks at high Reynolds numbers the power number is independent of the Reynolds number and hence of viscosity. However, the prediction of power consumption in gassed non-Newtonian fluids and the evaluation of gas dispersion have not so far been investigated.

We (1966) observed the power consumption in gassed non-Newtonian fluid comparing with Newtonian fluid. The experimental data carried out in glucamylase fermentation broth were first plotted as P_g/P_0 against N_a, where P_g/P_0 is the ratio of power consumption in gassed liquids to that in a non-gassed liquids at constant rotation speed and N_a is aeration number. For low-viscosity broths, there was very little effect of varying impeller diameter on the P_g/P_0 vs. N_a relationship. There had to be fairly high values of the Reynolds number (higher than 400) established in the vessel before the ratio of gassed to non-gassed power consumption dropped appreciably.

On the other hand, with the highly viscous broths, an appreciable reduction on the P_g/P_0 ratio was observed at Reynolds number of only 200. The real reasons for the difference in behavior between high- and low-viscosity broths are not known. In non-gassed, non-Newtonian fluids, mixing is often localized to a laminar region around the impeller at low agitation speeds. Aeration may expand this mixing region, resulting in increased power consumption. However, aeration certainly reduces bulk fluid density and viscosity, and hence may decrease the power consumption. The influences of aeration on the power consumption may be reduced in viscous fermentation broth by these counteracting effects. Nishikawa (1965) has studied power consumption in gassed C.M.C. solutions of differing rheological properties. He reported that, with differing viscosity indices, the flow pattern will change in the vicinity of rising bubbles and hence will influence the gassed power input. He also observed a considerable difference in gas holdup with change in viscosity index. Oyama and Endo (1955) reported that their data, when plotted P_g/P_0 vs. N_a, did not depend on Reynolds number. Rather, it was most sensitive to the geometry of the impeller. Their observations were limited to experiments with Newtonian systems. In the non-Newtonian systems, it was observed that the value of P_g/P_0 depends on the rheological properties of the system. The data were also plotted as P_g vs. $P_0^2 N D_i^3/Q^{0.56}$, according to the suggestion of Michel and Miller (1962). Fig. 8 is the plot for the non-Newtonian gluc-amylase fermentation broth. In the turbulent region, the correlation developed by Michel and Miller was found to be applicable even to a non-Newtonian fluid. The correlation could not be applied to non-Newtonian fluid in the laminar and transient region, i.e. low aeration number regions. This lack of correlation was particularly noticed with fluids of apparent viscosity greater than 300 c.p.

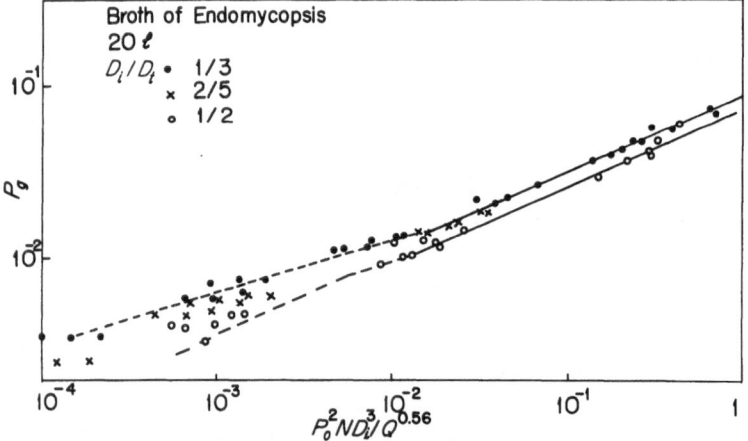

Fig. 8. P_g vs. $P_0^2 N D_i^3/Q^{0.56}$ for gluc-amylase fermentation broth

d) Motion of Gas Bubbles in Non-Newtonian Fluids

The velocity of the bubble in the gravitational field is related to the bubble volume by the following equation;

$$U = 84.0 \frac{v^{2/3}}{v} \qquad (6)$$

where U is bubble velocity, v is bubble volume and v is kinematic viscosity.

Oldshue (1966), by using the above equation, showed the various bubble sizes required to have various terminal velocities for fluids with a specific gravity of 1.0 and varying viscosities. If air bubbles are to rise from viscous fluids with any appreciable velocity, then they must have a reasonable size. For example in fluid of 10 poise viscosity, bubbles that are less than 0.76 mm in diameter will remain in the system, and only those that are 2.03 mm or larger will enter and pass out with any appreciable velocity. This, of course, causes much more difficulty in a small-scale fermenter than it would in a large-scale one.

Astarita and Gianin (1965) have reviewed the available theoretical knowledge of the motion of gas bubbles in Newtonian liquids, and studied the possibility of extension to non-Newtonian liquids.

In Stockes regime, the velocity-volume relationship is represented by the following equation,

$$U = \left[\frac{\varrho g}{C} \frac{2^{1+s}}{X_n} \left(\frac{4\pi}{3} \right)^{\frac{2-s}{3}} \right]^{\frac{1}{s}} v^{(1+s)/3s} \qquad (7)$$

where ϱ is specific gravity of liquid, C is consistency index, s is flow behaviour index and X_n is the coefficient defined by the following equation,

$$C_d = X_n/N_{Re} . \qquad (8)$$

Now, C_d is the drag coefficient and N_{Re} is a modified Reynolds number. It is worthwhile pointing out that for pseudoplastic fluids $(s < 1)$ the bubble velocity increases with increasing bubble volume more rapidly

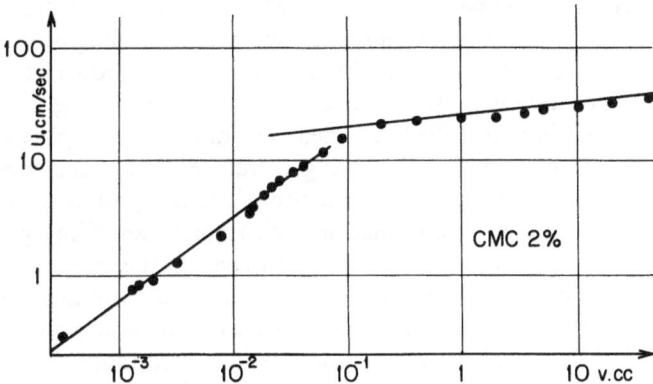

Fig. 9. Bubble velocities in a C.M.C. solution

than for Newtonian fluids as shown by Eq. (9).

$$\frac{d\log U}{d\log v} = \frac{1+s}{3s} > \frac{2}{3}.$$ (9)

A gas bubble moves in the Taylor regime when the Reynolds number is high, the bubble has a spherical cap shape, and the velocity-volume relationship is represented by the following equation.

$$U = 25.0v^{1/6}.$$ (10)

Fig. 9 is a plot of the terminal velocity vs. the bubble volume in the C.M.C. solution.

2. Shear Problems in Fermentation Fluids

The effects of agitation on microorganism morphology and biosynthesis are reported by many workers (Martin and Waters, 1952; Bartholomew, 1960). The performance of industrial fermentations is often determined by the type of agitation. Scale-up and fermenter operation are usually based on optimum aeration rate. However, it is important to consider the effect of shear in disrupting the living tissue which may be limiting in some fermentation processes, and to decide whether the shear effect is desirable or undersirable, depending upon the product under consideration. The solution of the shear problem may also aid in the design of impellers for gas dispersion, especially in viscous fermentation broths where the bubbles may not readily coalesce as described above.

Since mixing is turbulent in most aerobic fermentations, dynamic rather than viscous forces predominate, and disruption of living tissue no doubt occurs because of highly localized velocity gradients within or between eddies. Oldshue (1966) indicated that these local gradients of velocity in the vicinity of the impeller are orders of magnitude higher than the average shear rate in the stirred tank as shown in Fig. 10. However, one difficulty with observations made during an actual fermentation is that the effects of shear are not clearly separated from other effects of agitation, such as oxygen transfer. From this viewpoint, Midler and Finn's (1966) work on the use of model systems to study the effects of shear is helpful in eliminating such ambiguities of interpretation. They chose Tetrahymena, a kind of protozoa, as the model for shear studies. They observed that there is an additional slow decline in the fraction of survivors with time, the so-called "secondary" disruption, which follows the rapid "primary" damage in stirred tanks. The fraction of the cells surviving primary disruption varied with the diameter as well as with the rotation speed of the agitator. Fig. 11 shows that a plot

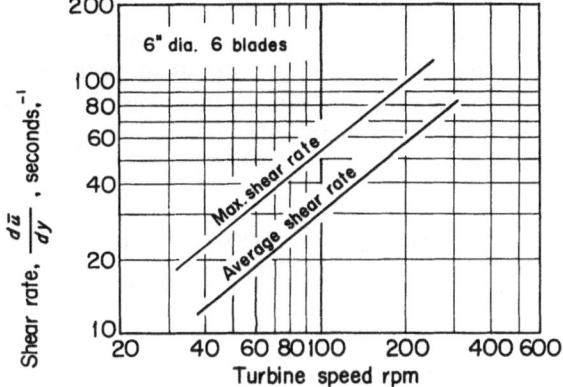

Fig. 10. Maximum and average shear rates in stirring tank

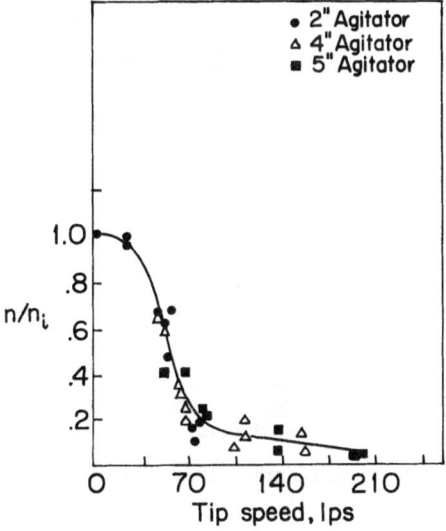

Fig. 11. Fractional survival vs. agitator tip speed

of survivors vs. tip speed of impeller correlated the data satisfactorily. On the other hand, neither the Reynolds number nor power input per unit volume seemed to be appropriate parameters.

Midler and Finn (1966) also studied disruption at relatively high shear rates in a Couette-type device. At shear rates below $1200 \, \text{sec}^{-1}$ only slight damage was observed. As shown in Fig. 12, data for two runs at the same viscosity of 69 c.p. and shear rate of $1970 \, \text{sec}^{-1}$, but at different gap widths, fall on the same line. Such consistency supports

2*

the assumption of defined laminar shear fields, and shows that the model system is responding to true shear effects. Increasing the viscosity of the cell suspension from 69 to 320 c.p. intensified both the primary and secondary damage at 1970 sec^{-1}. From these results they assumed that shear stress, rather than shear rate, is the important variable for disruption of living tissue.

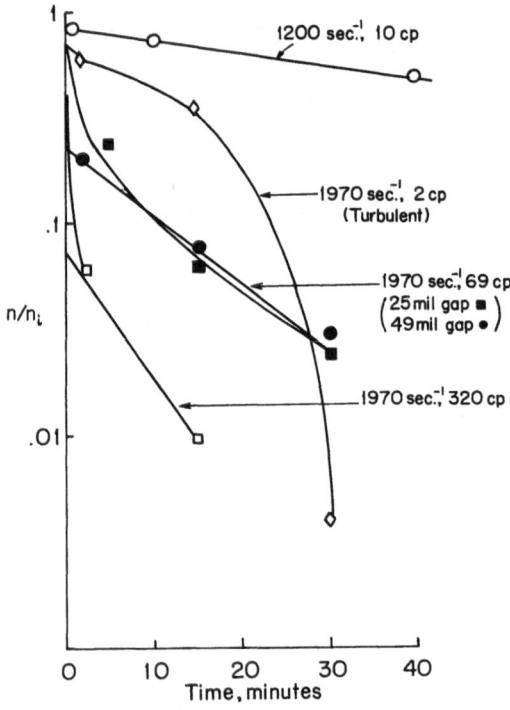

Fig. 12. *Tetrahymena* disruption in a laminar shear device

When the concept of the mean shear rate predicted by Metzner (1961) was applied, an agitator speed of 700 rpm used in Midler's experiment should have an effect equivalent to only 120 sec^{-1}. However, comparing this with the results in the Couette type device, the equivalent shear rate in those turbulent conditions should be 10–100 times those predicted by Metzner. Therefore, Midler and Finn suggested that maximum rather than average shear stresses must be considered in the presence of high local velocity gradients.

We (1968) also studied the influence of agitation on disruption of two kinds of mycelial pellets, *Lentinus edodes* and *Aspergillus niger*, in a conventional stirring tank. Since sometimes the growth type of pellet

reduces or eliminates its non-Newtonian behavior, it is also important
to combine these studies with the studies of the control of growth type
where product yield is concerned in fermentation processes. The
physical effects of agitation on the mycelial pellets were classified in two
types. One of these types is the decrease in diameter of the pellets due to
chipping off pellicles from their surface, and the other type is the direct
rupture of the spherical shape of pellets. In the latter case, the fraction

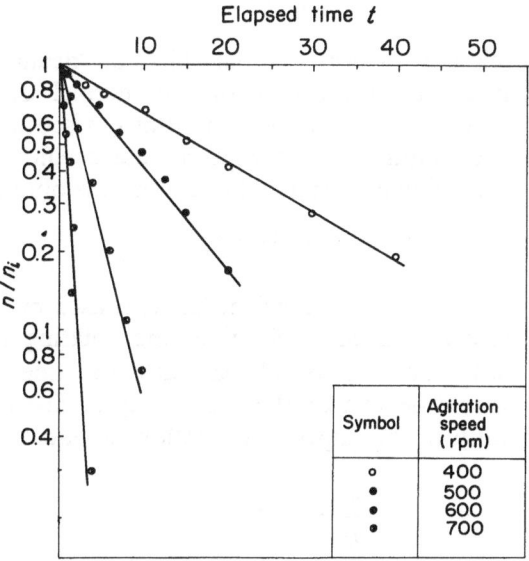

Fig. 13. Semilogarithmic plots of n/n_i vs. elapsed time using *Asp. niger*. Mean
diameter of pellets is 4.57 mm

of non-ruptured pellets with time can be plotted to give straight lines
on semilogarithmic coordinates as given in Fig. 13. The contributions
of stirring conditions and properties of mycelial pellets to the rupture
effects were analyzed quantitatively.

For the first of these effects the following experimental correlation
was found:

$$\frac{dD_p}{dt} = -k_c(ND_i)^{5.5} D_p^{5.7} \tag{11}$$

where D_i is impeller diameter, D_p is diameter of mycelial pellet and k_c is
experimental constant. In the first case, the decrease rate of pellet diameter varied with the diameter of the pellets in the stirring tank as well
as with the tip speed of the impeller.

In the second case, the rupture of mycelial pellets by agitation follows a monomolecular rate of reaction as shown in Eq. (12)

$$\frac{dn}{dt} = -k_r n \tag{12}$$

where n is the number of non-disrupted pellets, t is time and k_r is the rupture rate coefficient. Integrating under condition $n = n_i$ at $t = 0$,

$$n = n_i e^{-k_r t} . \tag{13}$$

The rupture correlation suggests random "hits" or "death by chance", somewhat similar to survivor curves for the thermal destruction of bacterial spores. Effects of agitation conditions, viscosity of liquid, pellet diameter and culture age on rupture rate coefficient, k_r, were studied, and the following experimental equation was obtained.

$$n = n_i e^{-\alpha(D_p^{3.2} N^{6.65} D_i^{8.72})t} . \tag{14}$$

It is clear that the rupture of mycelial pellets is caused by a high local stress, usually observed in the vicinity of the rotating impeller, or impeller shock itself. Since it is considered that the number of ruptured mycelial pellets should depend on the probability of circulation of the pellets in the vicinity of the impeller, the following equation could be introduced.

$$\frac{n}{n_i} = e^{-\beta(N D_i^3)t} \tag{15}$$

where β could be defined as an experimental constant determined by the correlation between the disruptive forces in the vicinity of the impeller and the resistance properties of mycelial pellets.

The tensile strength of specimens prepared from mycelial pellets of various fungi was measured by pendulum strength tester. The results are shown in Table 1. Both viscous stress and Reynolds stress need to be considered as shear stress to rupture mycelial pellets in a turbulent tank. The former is about 2×10^{-6} kg/cm^2, which is estimated using

Table 1. *The tensile strengths of the mycelial pellets*

Species	Tensile strength (kg/cm^2)
Asp. niger	0.278
Asp. oryzae	0.120
Lent. edodes	0.043
Ph. blakesleeanus	0.033

the maximum shear rate measured by Oldshue, and the maximum value of the latter is $0.046\,kg/cm^2$, which is estimated assuming that the maximum velocity fluctuations are half of that of the impeller tip speed under the condition that the impeller rotational speed is 900 rpm and the impeller diameter is 9 cm. Comparing these values with the tensile strength of mycelial pellets shown in Table 1, it is assumed that the viscous stress has little effect on the rupture of mycelial pellets. On the other hand, the maximum value of Reynolds stress is above the tensile strength of the mycelial pellet of *Lentinus edodes* and *Ph. blakesleeanus*, but below that of *A. niger* and *A. oryzae*. However, tensile strength

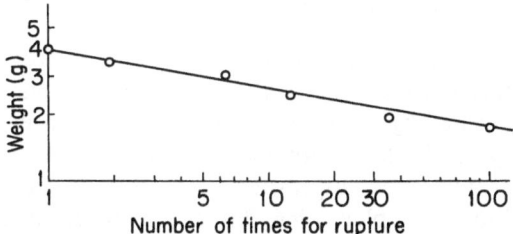

Fig. 14. Fatigue test against the specimen from mycelial pellet of *Asp. niger*

decreases with repeated loading, so a fatigue test of mycelial pellets of *A. niger* was carried out. By this fatigue test it was found that the number of times of repeating is proportional to the -5.77 power of the loading weight which is lower than the tensile strength of mycelial pellets, as shown Fig. 14. Eq. (15) may be rearranged as shown in Eq. (16) assuming that mycelial pellets should be ruptured with m times of circulation,

$$\frac{n}{n_i} = e^{-\frac{k}{mV}(ND_i^3)t} \tag{16}$$

where k is a constant.

The number of times of circulation required for disrupture of a pellet, m, is defined by the following equation:

$$m = \left(\frac{\tau_m}{\tau_0}\right)^{-5.77} \tag{17}$$

where τ_m is local maximum shear stress in a stirring tank and τ_0 is the tensile strength of mycelial pellet.

Combining Eq. (16) and Eq. (17), the following correlation can be introduced.

$$\frac{n}{n_i} = e^{-\left(\frac{k}{V}\right)\left(\frac{\tau_m}{\tau_0}\right)^{5.77}(ND_i^3)t} \tag{18}$$

If it is possible to assume that the maximum shear stress is proportional to the tip speed of the impeller, Eq. (18) has power index on the impeller diameter and rotation speed equivalent to those of Eq. (14).

Complete interpretation of the disruption of mycelial pellets is difficult. However, the results described above could mean that the strength of shear stress and the frequency are very important factors for the disruption of microbes in the stirring tank. For the purpose of this discussion, the development of a method for measuring the intensity of the turbulent fluctuations is necessary.

3. Physical Problems of Hydrocarbon Fermentations

Many studies on hydrocarbon as the carbon source for cultivating yeasts and bacteria to produce cellular proteins and various substances have been published by a number of investigators since 1960. Especially various findings on hydrocarbon fermentation such as pathway, cellular component, growth rate, respiration, and so forth have been reported (Humphrey, 1967; Sharpley, 1966; Miller and Johnson, 1966). However, many engineering problems remain to be solved before the microbial potential of hydrocarbon fermentation can be exploited on an industrial scale with an understanding of the physical properties of hydrocarbon fermentations.

In this section, some recent investigations carried out from the viewpoint of biochemical engineering are reviewed.

There are at least two troublesome physical problems that occur with hydrocarbon fermentation. These are the relative insolubility of hydrocarbon in water and the greater oxygen demand in a hydrocarbon fermentation.

Erickson, Humphrey, and Prokop (1969) have developed mathematical models which can be used to describe batch growth in hydrocarbon fermentation with two types of dispersed system. In the first type of fermentation, the growth-supporting hydrocarbon substrate is dissolved in inert hydrocarbons, and it is assumed that substrate utilization from the dispersed phase causes little or no change in the interfacial area.

The dewaxing of crude oil by microbes is one example of this fermentation. In the models, the possibilities of growth occuring at the surface of the dispersed phase and in the continuous phase are considered. Therefore three special cases were examined, as follows; (1) all growth occurs at the surface of the dispersed phase; (2) growth occurs both at the interface and in the continuous phase, and the substrate equilibrium is continuously established between the two phases; (3) growth occurs both at the interface and in the continuous phase, and the substrate consumption in the continuous phase is limited by rate of transport of substrate to that phase. Consequently, the comparison of the first case model with available experimental data showed good agreement between model and data.

Let N_d be number of cells required to cover the surface of one drop, Y be yield in number of cells per unit mass of substrate, S_0 be initial concentration of substrate in dispersed phase, and V_d be volume of dispersed drop, then the value of the dimensionless parameter N_d/YS_0V_d plays an important part in determining the growth curve decided by the model described above (cf. Fig. 15).

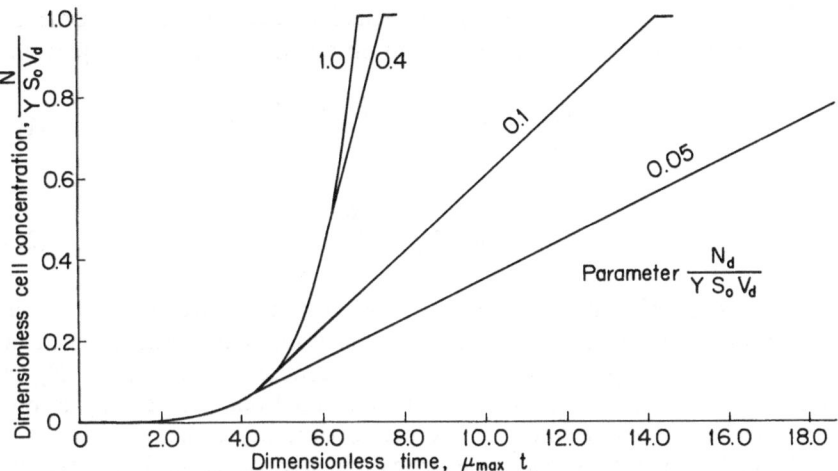

Fig. 15. Batch growth curve for cells grown on a single hydrocarbon drop

In the second type of system, the dispersed phase is assumed to be pure substrate; therefore, the substrate concentration in the dispersed phase remains constant but the interfacial area decreases as substrate is consumed. The growth of yeast on n-alkanes is one example of this type.

In this type of dispersed system some of the cells will be adsorbed at the drop surface and the others will be suspended in the continuous phase also. However if the dispersed phase is pure substrate, those cells attached to the drop surface will have a specific growth rate equal to the maximum specific growth rate, and those cells suspended in the continuous phase will have a specific growth rate that can be represented by Monod's equation. When mass transfer is important, the rate of growth in the continuous phase is limited by the quantity of substrate that diffuses into that phase. Since the growth at the drop

surface and the mass-transport limited growth are both directly related
to the interfacial area, it is important to consider how the interfacial
area might change during the course of a batch fermentation.

Erickson and Humphrey (1969) used those mathematical models
which have been used to describe batch growth in hydrocarbon
fermentation to predict the behavior of continuous hydrocarbon
fermentation in a chemostat.

Aiba, Moritz, Someya, and Haung (1969) also recognized that a
favorable contact in the cultivation between cells and n-alkane droplets
seemed indispensable to produce a normal pattern of yeast growth.
They studied yeast growth kinetics in a fermenter, coupled with an
empirical correlation between the interfacial area per unit volume of
liquid and the operation condition, and obtained the following equation.

$$\mu = \frac{\mu_{max}}{K_s} \gamma' N^{1.2} D_i^{0.3} \left(\frac{H_L}{D_t}\right)^{-1.2} \phi^{0.24} \tag{19}$$

provided:

$$\gamma' = 1.7 \gamma \left(\frac{\varrho}{\sigma}\right)^{0.6} \tag{20}$$

where μ is specific growth rate at the logarithmic growth phase, μ_{max} is
maximum value of μ, K_s is saturation constant, N is rotation speed of
impeller, D_i is impeller diameter, D_t is diameter of fermenter, H_L is
liquid depth, ϕ is volume fraction of oil in medium, ϱ is density of oil,
σ is interfacial tension and γ, γ' are proportionality constants.

In order to obtain favorable growth in hydrocarbon fermentation,
the substrate must be emulsified into the aqueous phase and to bring
enough microorganism into contact with the substrate. It has been
suggested that the hydrocarbon dissolved in continuous phase is not
the main source of substrate, but rather that uptake of hydrocarbon
occurs by attaching microorganisms to droplets of hydrocarbon as
described above. In the process of attachment, it is probable that
microorganisms have strong affinity to the hydrocarbon. The affinity
with microorganisms is the most essential factor in the emulsification
of hydrocarbon into a continuous phase.

Many studies have been reported about the methods of emulsification
of hydrocarbons, such as emulsification by surface-active agents, by
mechanical agitation, or by biological methods. For the most favorable
configurations between oil droplets and cells for growth, the "bio-
logically active emulsion of hydrocarbon" was proposed. Munk (1969)
indicated that the larger the amount of lipid a cell contained, the faster
it incorporated hydrocarbon. Mimura et al. (1970) studied the characteristic phenomena of emulsification in hydrocarbon fermentation, and

found that hydrocarbon-assimilating yeasts have a stronger affinity for oil than non-assimilating yeasts. They extracted lipophilic compounds, which have excellent activities in emulsification of hydrocarbon and water, from *Candida* sp. cells. Most synthetic surface-active agents weakened the affinity of the cell for hydrocarbon, however, the naturally occurring surface-active compounds of biological origin promoted the growth of the yeast.

The other important physical problem with hydrocarbon fermentations is the increased oxygen demand. The very high oxygen demand per unit of cell in hydrocarbon fermentation has been pointed out by several investigators, including Johnson (1964) and Darlington (1964). For example, it is easy to estimate from typical yeast cell analyses that almost three times as much oxygen is required in a hydrocarbon fermentation. Though there have been a number of reports dealing with the problems of agitation and aeration in the usual carbohydrate fermentation, few reports are concerned with the agitation and aeration in hydrocarbon fermentation. Mimura *et al.* (1969) studied the effect of hydrocarbon oil on oxygen transfer rate coefficient in hydrocarbon fermentation, and found that the coefficient increased exponentially with the addition of kerosene oil. From shaken flask experiments, an empirical equation which shows the oxygen transfer rate coefficient in the oil-water system is obtained as follows;

$$\frac{K_L a}{H} = 54\, V_a^{-0.56} + 10^{-6} \exp 3.3 \frac{V_k}{V_t} \tag{21}$$

where H is Henry's constant, V_a is volume of aqueous solution in flask, V_k is volume of kerosene in flask and V_t is total volume of liquid.

For the purpose of process design for hydrocarbon fermentation, it is important to know how the oxygen transfer in a culture system containing hydrocarbon differs from the usual carbohydrate fermentation.

Concluding Remarks

It has been possible here to deal only briefly with the correlations between the physical properties of fermentation fluids and the operating conditions in the fermenter found by many investigators. For example, if it were known what was required in the way of shearing action at each step of the fermentation process, then it would be possible to carry out a relatively rigorous calculation throughout the system. Thus, geometric non-similarity in scale-up of the fermenter could be applied to reduced or increased fluid shear stress.

There are many other investigations which have not been described at all in this review, and of which little mention has been made, such as foaming problems in fermentation processes.

The diversity of fluid characteristics is so great from one micro-organism to another that little generalization is possible. Moreover the products and kinds of fermentation processes are likely to expand, particularly the use of hydrocarbons and natural gas for biosynthesis. Therefore, much more work is needed to determine the influence of fermentation fluid characteristics on the fermentation process and product recovery. It needs not only studies of superior and more quantitative measurements of the physical properties of cells, such as cell density, cell size, the electrostatic property of the cell surface, specific heat and heat conductivity of cells, but also on the properties of cell suspensions, such as morphology, viscosity and surface tension, to obtain the fundamental data for practical equipment design and to control unit operations in the fermentation industry, such as mechanical separation and disintegration of cells for product recovery, medium sterilization, air sterilization, control of dispersion, as well as the actual fermentation process.

In passing, let us say that a specific aim of this work is to demonstrate that a biochemical engineering approach to the microbial world deserves due attention.

Nomenclature

A	empirical constant
C	consistency
C_d	drag coefficient
D_i	impeller diameter
D_p	diameter of mycelial pellet
D_t	vessel diameter
E	activation energy
g	acceleration due to gravity
H	Henry's constant
H_L	liquid depth
i	number of impeller stages
$K_L a$	volumetric coefficient
K_s	saturation constant
k	empirical constant
k_c	empirical constant
k_r	empirical constant
L	blade length of impeller
M_f	momentum factor
m	number of times of circulation for disruption of a pellet
N	rotation speed of impeller
N_d	number of cells required to cover the surface of a hydrocarbon drop
N_{Re}	Reynolds number

n	number of non-disrupted pellets
n_i	initial number of pellets
P_g	power consumption in gassed system
P_0	power consumption in ungassed system
Q	aeration rate
R	gas constant
S_i	impeller spacing
S_0	initial substrate concentration
s	flow behavior index
T	temperature
t	time
t_m	mixing time
U	bubble velocity
V	volume of liquid
V_a	volume of aqueous solution in flask
V_d	volume of dispersed drop
V_k	volume of kerosene in flask
V_t	volume of liquid in flask
v	bubble volume
W	blade width of impeller
X_n	coefficient defined by Eq. (8)
Y	yield constant

Greek Letters

α	empirical constant
β	empirical constant
γ	proportionality constant
γ'	proportionality constant
η	viscosity
η'	plastic viscosity
μ	specific growth rate
μ_{max}	maximum value of μ
ϱ	density of liquid (continuous phase)
σ	interfacial tension
ν	kinematic viscosity
τ_m	local maximum shear stress in an agitation tank
τ_0	tensile strength of mycelial pellet
ϕ	volume fraction of oil in medium

References

Aiba, S., Moritz, V., Someya, J., Haung, K. L.: J. Ferment. Technol. 47, 203 (1969).
— Haung, K. L., Moritz, V., Someya, J.: J. Ferment. Technol. 47, 211 (1969).
Andrade, E. N.: Nature 125, 309 (1930).
Astarita, G., Apuzzo, G.: A. I. Ch. E. Journal 11, 815 (1965).
Bartholomew, W. H.: Advance in Appl. Microbiol. 2, 289 (1960).
Blakebrough, N., Sambamurthy, K.: Biotechnol. Bioeng. 8, 25 (1966).

Chain, E. B., Gualandi, G., Morisi, G.: Biotechnol. Bioeng. **8**, 595 (1966).
Darlington, W. A.: Biotechnol. Bioeng. **6**, 241 (1964).
Deindoerfer, F. H., Wise, J. M.: Advance in Appl. Microbiol. **2**, 265 (1960).
Erickson, L. E., Humphrey, A. E., Prokop, A.: Biotechnol. Bioeng. **11**, 449 (1969).
— — Biotechnol. Bioeng. **11**, 467 (1969).
— — Biotechnol. Bioeng. **11**, 489 (1969).
Fondy, P. L., Bates, R. L.: Paper presented at Am. Ins. Chem. Eng.-C.I.C. Meeting, Cleveland, May (1961).
Humphrey, A. E.: Paper presented at Am. Ins. Chem. Eng.-C.I.C. Meeting, Cleveland, May **9**, 3 (1967).
Johnson, M. J.: Chem. Ind. London **36**, 1532 (1964).
Martin, S. M., Waters, W. R.: Ind. Eng. Chem. **44**, 2229 (1952).
Metzner, A. B., Taylor, J. S.: A. I. Ch. E. Journal **6**, 109 (1960).
— Feehs, R. H., Ramos, H. L., Otto, R. E., Tuthill, J. D.: A. I. Ch. E. Journal **7**, 3 (1961).
Michel, B. J., Miller, S. A.: A. I. Ch. E. Journal **8**, 262 (1962).
Midler, M., Jr., Finn, R. K.: Biotechnol. Bioeng. **8**, 71 (1966).
Miller, T. L., Johnson, M. J.: Biotechnol. Bioeng. **8**, 549 (1966).
Mimura, A., Kawano, T., Kodaira, R.: J. Ferment. Technol. **47**, 229 (1969).
— Watanabe, S., Takeda, I.: J. Ferment. Technol., in press (1970).
Munk, V., Dostalek, M., Volfova, O.: Biotechnol. Bioeng. **11**, 383 (1969).
Nishikawa, F.: M. S. Thesis, Univ. of Pennsylvania (1965).
Ohyama, Y., Endoh, K.: Chem. Eng. (Japan) **19**, 2 (1955).
Oldshue, J. Y.: Biotechnol. Bioeng. **8**, 3 (1966).
Owen, S. P., Marvin, J., Johnson, M. J.: Appld. Microbiol. **3**, 375 (1955).
Phillips, D. H., Johnson, M. J.: J. Biochem. Microbiol. Tech. Eng. **3**, 277 (1961).
Richards, J. W.: Prog. in Ind. Microbiol. **3**, 141 (1961).
— British Chem. Eng. **8**, 158 (1963).
Satoh, K.: J. Ferment. Technol. **39**, 517 (1961).
— J. Ferment. Technol. **41**, 588 (1963).
Sharpley, J. M.: Elementary Petroleum Microbiology, p. 91. Houston: Gulf Pub. Corp. (1966).
Steel, R., Maxon, W. D.: Biotechnol. Bioeng. **4**, 231 (1962).
— — Biotechnol. Bioeng. **8**, 97 (1966).
— — Biotechnol. Bioeng. **8**, 109 (1966).
Taguchi, H., Miyamoto, S.: Biotechnol. Bioeng. **8**, 43 (1966).
— Kimura, T.: J. Ferment. Technol. **48**, 117 (1970).
— Yoshida, T.: J. Ferment. Technol. **46**, 814 (1968).
Takeda, K., Hoshino, T., Taguchi, H., Fujii, T.: Chem. Eng. (Japan) **32**, 376 (1968).

H. Taguchi
Dept. of Fermentation Technology
Faculty of Engineering
Osaka University, Osaka

CHAPTER 2

Separation of Cells from Culture Media

SHUICHI AIBA and MASAHARU NAGATANI

With 11 Figures

Contents

1. Introduction

Various equipment and procedures used for the separation of cells from culture media in the biochemical industries are similar, in principle, to those used in the chemical industries. No marked differences are noted among filter press, centrifuge, decantor and so forth now being used in both industries. It seems significant to point out, however, that some aspect of the cell separation in the biochemical industries differs essentially from the separation technique of the chemical industries.

Needless to say, the cellular material – yeast, bacterium, fungus, actinomycete and virus – is different from solid, liquid and gas phases, a combination of the three phases being employed to specify the separation technique in the chemical industries. In addition to the cellular proximity to water in density, the size ranges from microns to millimicrons. The shape of fungus and/or actinomycete cannot be defined *in situ*, being far from spherical and/or ellipsoidal. Cellular content in the fermentation mash or in an aqueous suspension is usually less than several per cent. The cell separation from the dilute suspension requires a variety of operations which are not needed, otherwise.

Owing primarily to the organic character of the cell and the rich content in water, the suspension, once concentrated, tends to exhibit the non-Newtonian behavior. Transportation, heat transfer and mass transfer (i.e. drying and so on), all of which are the subjects adjunct to the separation, abound with problems still remaining to be dissolved in regard to the non-Newtonian characteristics.

One of the points to feature the cell separation is that the filtrate or effluent must usually be free from undue damages on flavor and taste as experienced in the brewing industry and/or must be least in number concentration of coliform bacteria as required in the sedimentation of activated sludge in the biological treatment of sewage.

Cells are liable to lysis under harsh conditions in temperature, pH and in pressure. Cellular cakes in the filtration are unexceptionally compressible. Some points mentioned above require particular attention which is not demanded of the separation in the chemical industries.

. Flocculation and deflocculation, followed by sedimentation under gravity will be the subjects of discussion in this chapter. Even though the microorganisms due to be treated here in each unit operation will be different and inconsistent from one to another, depending on the subject, the aspect underlying the discussion will be made consistent such that the difference, if any, of the separation will be silhouetted between the biochemical and chemical industries. Because of the idea not to include especially the filtration and consolidation in this chapter, some brief comment on the recent advance in connection with these separation techniques will be needed.

By and large, the filtration of fungus, especially actinomycete broth in the fermentation industry has been the impediment to a rationale of plant operation. In some instances, however, a specific technique of heating the broth, causing so far a certain extent of coagulation of the cellular protein has been found effective in decreasing the specific resistance of the filter cake. Another technique of having actinomycete autolyzed prior to the filtration decreased also the cake resistance. Both techniques are being used on an industrial scale [14, 15].

The compressibility of filter cake composed of yeast, bacterium, fungus or actinomycete or the combination characterizes, indeed, the filtration here. The compressibility factor assigned to the exponent of the pressure difference or the driving force in filtration is needed for a proper design and operation. Some experiment to assess the exponent and the representative values have recently been demonstrated [10, 11].

In regard to filtration, recent advances in the brewing industries must also be referred to briefly. Absolute filters such as "Millipore" in cartridges are now in use for the cell separation from the brewery product to avoid the after effect of heat sterilization.

By consolidation it is meant that liquid (beverage, Japanese saké, soy-bean sauce and so on) is squeezed out by means of either hydraulic or pneumatic pressure. The problem whether such squeezing to minimize the liquid content in the cake to an extent of 50% or less is really worthwhile or not remains to be solved.

Engineering problems involved in the consolidation will require some sophisticated approach to the solutions, because the compressibility exponent of the cake and the ratio of dry to wet cake vary depending on the progress of consolidation [9]. A specific device now in active use in Japan in the consolidation of "saké" mash will be referred to, if needed [6].

2. Observation of Flocculation and Deflocculation

2.1. Experimental Apparatus [4]

An optical arrangement to observe continuously the flocculation and deflocculation phenomena of microbial cells and/or particulate particles suspended in liquid in an agitated vessel is shown in Fig. 1a. A principal part of the apparatus is a cylindrical vessel of acrylic resin (nominal volume $= 780\ cm^3$, working volume $= 700\ cm^3$). A paddle impeller of stainless steel agitates the suspension.

A sensor of cadmium sulfide to detect continuously the turbidity of the suspension in the vessel is shown in the figure by the two rod-like bodies. The construction details of the light source and the light recipient are shown in Fig. 1b and 1c.

The geometrical configuration of the light source and the recipient remains fixed. The counterparts at the bottom and the lid of the vessel are co-axial and the distance between both ends of the parts is adjusted to 10 mm. The measurement here deals with the microbial cells and its flocs passing incidentally through ,the definite interval.

When the floc-size determination is required, the cell suspension is bypassed into a glass sampler of flat Pyrex plates (Fig. 1d) to take photomicrographs of the suspension at standstill.

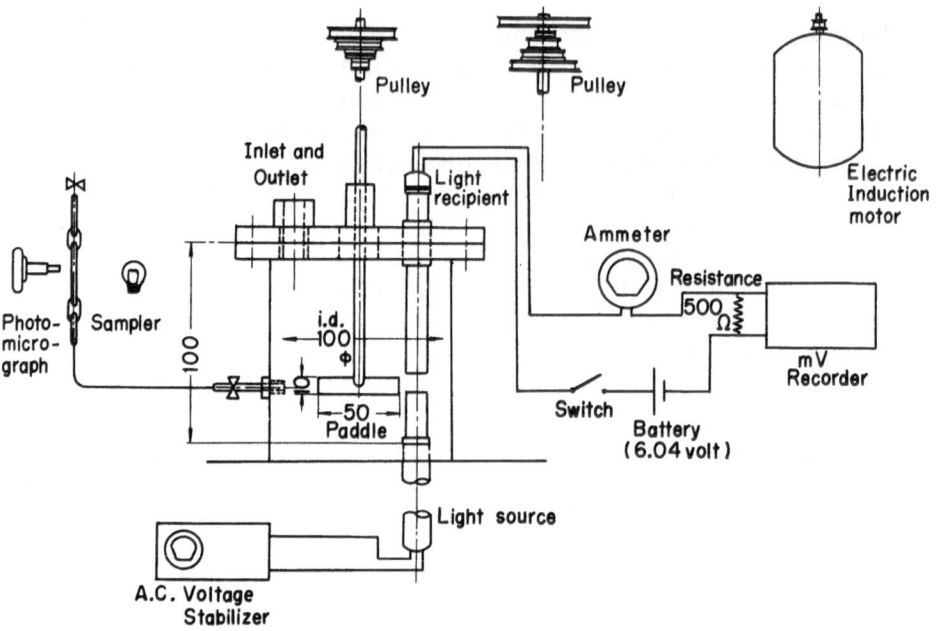

Fig. 1a. Optical arrangement to observe continuously flocculation and deflocculation

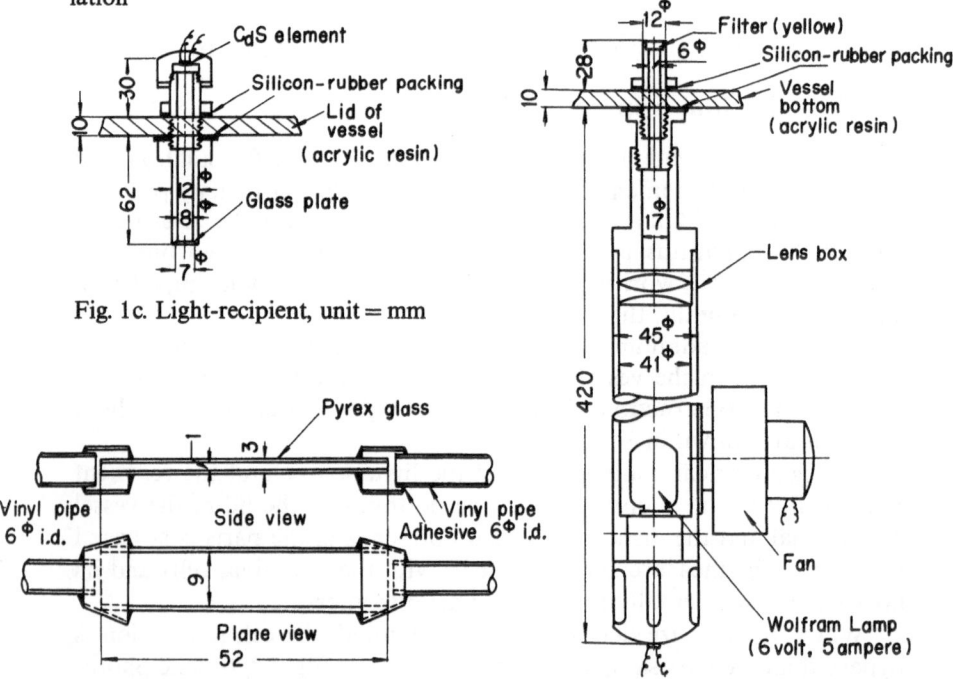

Fig. 1c. Light-recipient, unit = mm

Fig. 1d. Sampling device, unit = mm

Fig. 1b. Light-source, unit = mm

Assuming that a definite resistance (500 ohm) in the circuit is small enough, the resistance, R of the CdS element is correlated with the recorder reading, V (millivolt) by the following equation.

$$R = \frac{6.04 \times 0.5}{V \times 10^{-3}}$$

$$= \frac{3.02}{V} \times 10^3 \text{ kiloohm} .$$

(1)

The characteristic value, δ of the CdS element is defined as follows:

$$\delta = \frac{\log(R_1/R_2)}{-\log T_1 - (-\log T_2)}$$

$$= \frac{\log(R_1/R_2)}{\log(I_0/I_1) - \log(I_0/I_2)} .$$

(2)

The value of δ is given as a slope of a correlation in Fig. 2 between the value of R and the turbidity, $-\log T$ with respect to an aqueous suspension of *Chlorella* cells (see later). The value of δ is least susceptible to the changes in light intensity of the wolfram lamp and in temperature.

Suppose the cell suspension be subjected to flocculation in the agitated vessel to observe a transfer of the recorder reading, V millivolt from the initial value ($V_1 = 15\,\text{mV}$); an increase of V implies the progress of flocculation because of an increase of the incident light onto the CdS element and *vice versa*. A new reading, $V = V_2$ can be converted to the turbidity, $-\log T_2$ by calculating the value of R_2 from Eq. (1) and by substituting the R_2 value into Eq. (2) (cf. solid line in Fig. 2).

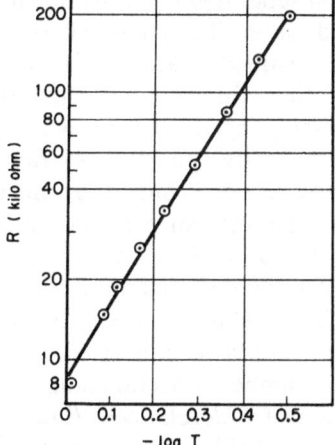

Fig. 2. R vs. $-\log T$ (calibration with *Chlorella* cells)

3*

2.2. Observation Procedure [4]

The microbial cells used are *Chlorella vulgaris* Beijerink C-30. For compositions of the culture medium, see Table 1. An inoculum into a drum type of glass vessel (working volume = 800 cm^3) was shaken under a fluorescent lamp, 3,000 to 4,000 luxes at 20° C for about two weeks.

Table 1. *Medium Composition (Chlorella vulgaris)*

glucose	5.0	g
$Ca(NO_3)_2 \cdot 4 H_2O$	1.0	g
$MgSO_4 \cdot 7 H_2O$	0.25	g
KH_2PO_4	0.25	g
KCl	0.25	g
Fe solution [a]	1.0	ml
A_5 solution [b]	1.0	ml
distilled water	1,000	ml
pH = 6		

[a] Composition of Fe solution.

$FeSO_4 \cdot 7 H_2O$	2.0 g
conc. H_2SO_4	4 drops
distilled water	1,000 ml

[b] Composition of A_5 solution.

H_3BO_3	2.9	g
$MnCl_2 \cdot 4 H_2O$	1.8	g
$ZnSO_4 \cdot 7 H_2O$	0.2	g
MoO_3	0.02	g
$CuSO_4 \cdot 5 H_2O$	0.08	g
distilled water	1,000	ml

The cells harvested and separated from the medium with a centrifuge were rinsed with distilled water. Then the cells were suspended in an appropriate amount of distilled water such that the turbidity at 530 millimicrons showed the value of 0.5 in this preparation. Without any addition of a dispersing agent the suspension was found well-dispersed. Cells used were nearly spherical, average diameter, $d_p = 3.2$ microns, while the density, ϱ_p determined by measuring the rate of interfacial subsidence of the cell suspension under gravity and by applying the Stokes' law was $\varrho_p = 1.09$ g/cm^3 at 20° C.

Regarding the value of $-\log T$ vs. cell mass concentration, X a linear relationship was established. Values of $X = 0.255$ mg dry cell/ml which correspond to $-\log T = 0.5$ could be converted *via* the counting with a hemacytometer to the cell number concentration, $N = 4.1 \times 10^7$ cells/ml.

The aqueous suspension of *Chlorella* cells (initial turbidity, $-\log T_1 = 0.5$) is charged into the vessel (Fig. 1) and the impeller is rotated. Due to the greenish appearance of the cells a yellow filter is used at the end of the light source (cf. Fig. 1a). Confirming that the recorder reading

($V_1 = 15$ millivolt) remains unchanged for about 5 min, a small amount of metallic ions, either aluminum chloride or calcium chloride is added into the suspension to initiate the flocculation.

Time, $t = 0$ at which either flocculation or deflocculation starts is clearly observed from the recorder. Each phenomenon is assumed steady, if a final reading, V_2 of the recorder is kept unchanged for nearly 20 min.

2.3. Reversibility between Flocculation and Deflocculation

Fig. 3 demonstrates a reversibility between flocculation and deflocculation observed with the *Chlorella* suspension by changing the rotation

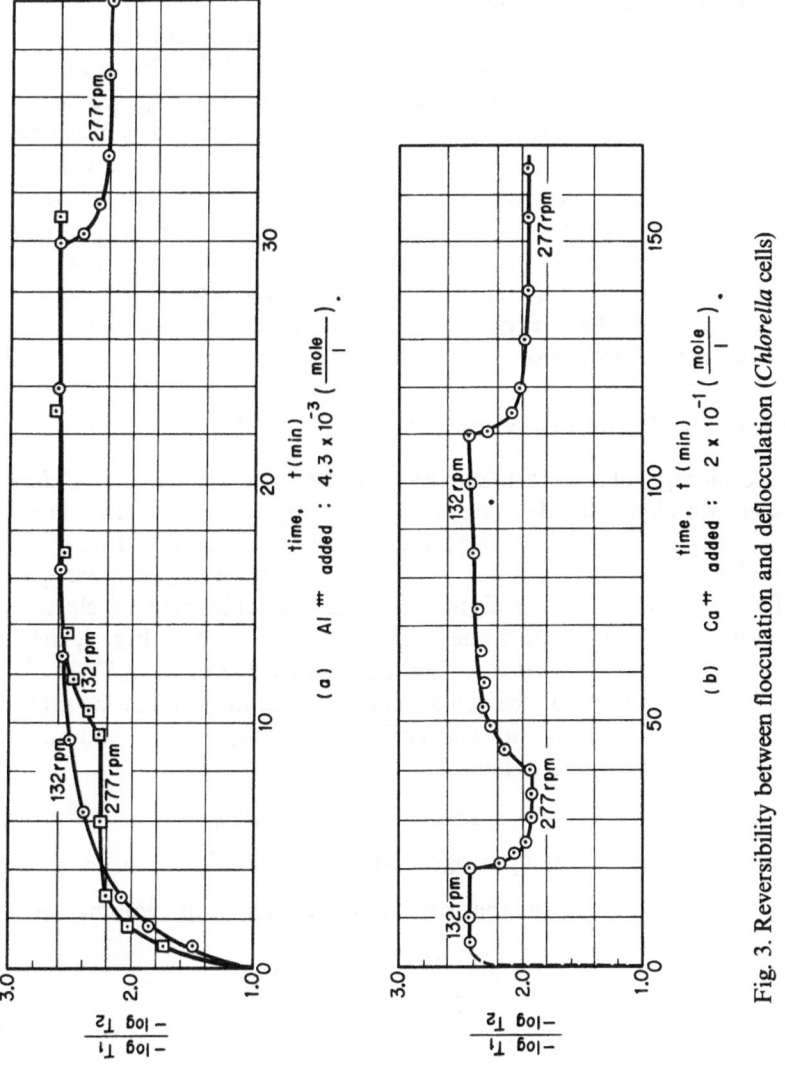

Fig. 3. Reversibility between flocculation and deflocculation (*Chlorella* cells)

speed of the impeller from 132 rpm to 277 rpm and *vice versa*. Aluminum chloride was used in Fig. 3 (a), while calcium chloride was used as an electrolyte in (b). Each run was at room temperature and the pH value of the suspension ranged from 4.0 to 4.5.

The increase and decrease of $-\log T_1/-\log T_2$ values with time, t in Fig. 3 indicate the progress of flocculation and deflocculation, because the increase of $-\log T_1/-\log T_2$ implies the decrease of $-\log T_2$ and *vice versa* due to the fixed value of $-\log T_1 = 0.5$. It is interesting to point out from Fig. 3 that the values of $-\log T_1/-\log T_2$ in these diagrams reach the same level, irrespective of the previous condition of agitation, if the rotation speed of the impeller is fixed. Briefly, no hysteresis is noted.

Reich *et al.* discussed the reversibility by using carbon and ferric oxide flocs formed in water under agitation [13]. The reversibility referred to by them, however, was not so evident, compared with the data in Fig. 3. The flocculation of *Chlorella* cells originates presumably from the alleged bridge formation among the individual cells. The bridge must have been constructed principally by the polyhydric and metallic ions added into the suspension. Though the assumption on the flocculation mechanism was derived referring exclusively to the experimental data (Fig. 3), the microbial cells suspended in liquid (water) and its flocs in general are expected to belong in the same category as this specific demonstration of the *Chlorella* cells.

2.4. Water Content of Microbial Floc

Photomicrographs were taken by sampling (cf. Fig. 1d) the *Chlorella* suspension at the steady state of V_2 (or $-\log T_2$) to correlate the value of $-\log T_2$ with the mean surface diameter, $\bar{d}_f$ of the floc. Fig. 4 shows the plot of $-\log T_1/-\log T_2$ against $\bar{d}_f/d_p$ in a logarithmic paper. Different symbols in the figure indicate the difference in electrolyte conditions used. If a solid line is assumed as shown in Fig. 4, the floc size, $\bar{d}_f$ could be assessed from the observed value of $-\log T_2$.

Taking the fact that the cell suspension is dilute enough for granted, a relationship between the suspension turbidity, $-\log T$ and the cell number concentration, N is:

$$-\log T_1 = \log(I_0/I_1) = k_1 (\pi/4) \, d_p^2 N_1 , \qquad (3)$$

$$-\log T_2 = \log(I_0/I_2) = k_2 (\pi/4) \, \bar{d}_f^2 N_2 . \qquad (4)$$

A mass balance of the cell material per unit volume of the suspension is:

$$(1 - \varepsilon) \frac{\pi}{6} \bar{d}_f^3 N_2 = \frac{\pi}{6} d_p^3 N_1 . \qquad (5)$$

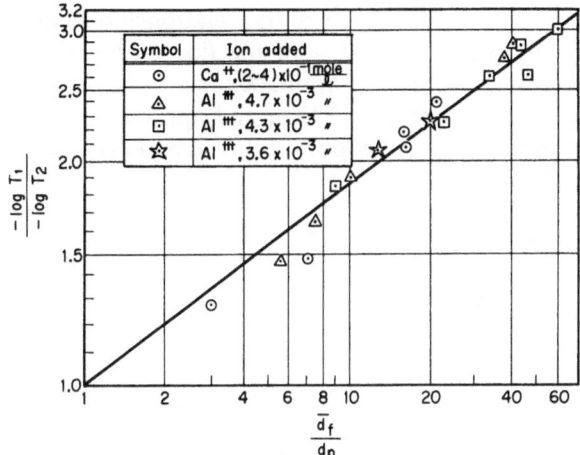

Fig. 4. $-\log T_1/-\log T_2$ vs. $\bar{d}_f/d_p$ (*Chlorella* cells)

From Eqs. (3), (4), and (5),

$$\frac{-\log T_1}{-\log T_2} = (k_1/k_2)\,(1-\varepsilon)\,(\bar{d}_f/d_p)\,. \tag{6}$$

The proportionality constants, k_1 and k_2 in Eqs. (3) and (4) must have the implication of adsorption coefficient which is dependent on the value of d_p and $\bar{d}_f$ [22].

Assuming that the floc in question is a loose aggregate of the individual cells, entrapping a considerable amount (fraction) of water in the intracellular space, the values of k_1 and k_2 in this example are deemed in connection with each cellular particle, even though the value of $\bar{d}_f$ is larger considerably than the original size, d_p.

Accordingly, it is assumed that $k_1 = k_2$. Assuming also that,

$$1-\varepsilon = (\bar{d}_f/d_p)^m\,.$$

Eq. (6) is rearranged as follows:

$$-\log T_1/-\log T_2 = (\bar{d}_f/d_p)^{m+1}\,. \tag{7}$$

The exponent, $(m+1)$ in Eq. (7) can be estimated from the slope of the solid line in Fig. 4 as shown below.

$$m+1 = 0.267\,,$$
$$m = -0.733\,.$$

Then the cell content, $(1-\varepsilon)$ of the floc varies, depending on its size, as follows:

$$1-\varepsilon = (\bar{d}_f/d_p)^{-0.733}\,. \tag{8}$$

A solid curve (Eq. (8)) in Fig. 5 is the relationship between $(1 - \varepsilon)$ and $\bar{d}_f$. Clearly the water content, ε of the floc increases appreciably till the size increases to an extent of about 20 times as large as the average diameter of single cells, levelling off then and approaching the value of 95%. No experimental data to support and/or denounce Eq. (8) directly has ever been published.

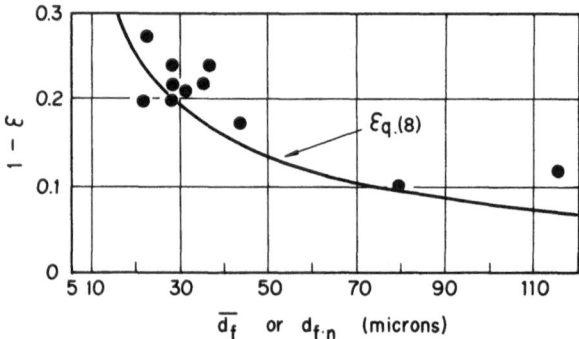

Fig. 5. $1 - \varepsilon$ vs. $\bar{d}_f$ or $d_{f \cdot n}$

However, Mueller et al. presented recently the data on the nominal diameter of activated sludge and the water content [8]. Assuming the spherical floc, the nominal diameter, $d_{f \cdot n}$ microns is determined from the observations of dry and wet densities of the floc, the number of flocs per unit volume of liquid and in addition, from the value of MLSS (mixed liquor suspended solids).

The solid circles in Fig. 5 are the data, $(1 - \varepsilon)$ vs. $d_{f \cdot n}$, cited from the work of Mueller et al. Although the definition of the floc diameter and the procedure of assessing the water content are different and difficult to compare between the works of Aiba et al. and Mueller et al., the similar pattern of $(1 - \varepsilon)$ as affected by the floc size is still interesting.

Tambo et al. measured the density of aluminum floc by observing the floc size and the settling velocity under gravity [19]. Reinterpreting the results in the light of Eq. (8), their empirical formula could be represented by the following equation.

$$1 - \varepsilon = \frac{0.0003}{(\varrho_s - 1) \, d_{f \cdot v}^{1.30 \sim 1.45}} \, . \tag{9}$$

Eq. (9) is valid for $d_{f \cdot v}$ ranging nearly from 200 to 2,000 microns. Taking $\varrho_s = 2.0 \text{ g/cm}^3$,

$$1 - \varepsilon = \frac{0.0003}{d_{f \cdot v}^{1.30 \sim 1.45}} \, . \tag{10}$$

The basis of floc-size assessment in Eq. (10) is also different from the previous examples. The value of $(1 - \varepsilon)$ with respect to $d_{f \cdot v} = 180$ microns, for example, ranges from 0.05 to 0.10, if Eq. (10) is used.

Another rough estimate of $(1 - \varepsilon)$ with reference to an empirical equation presented by Thomas on the various kinds of flocs of inorganic materials ranging from 100 to 3,000 microns in size [20] gives $(1 - \varepsilon) = 0.13$ commensurate to $\bar{d}_f = 180$ in Fig. 5.

3. Some Analysis of Flocculation and Deflocculation

3.1. Reaction-Rate Constant of Flocculation [5, 21]

The rate of change in number concentration, N_k of specific flocs, each of which is composed of k single cells can be expressed by [5]:

$$\frac{dN_k}{dt} = \frac{1}{2} \sum_{i=k-j}^{j=k-1} K_{ij} N_i N_j - \sum_{j=1}^{\infty} K_{kj} N_k N_j. \tag{11}$$

A well-dispersed suspension, the turbidity of which is shown by $-\log T_1$ be subjected to flocculation to show the change of turbidity, $\Delta(-\log T)$ for Δt after the initiation of the flocculation.

Since the value of Δt is small, the flocculation deals primarily with the formation of specific flocs of two cells. The decrease of single cell number and the increase of the flocs of the two cells in the suspension will be discussed.

A decrease of number concentration, ΔN_1 of single cells during Δt will be:

$$\Delta N = -K_{11} N_1^2 \Delta t. \tag{12}$$

On the other hand, an increase of number concentration, ΔN_2 of the flocs comprising the two cells during the same period, Δt is:

$$\Delta N_2 = (1/2) K_{11} N_1^2 \Delta t. \tag{13}$$

The turbidity, $-\log T$ of the suspension after Δt,

$$-\log T = k_1 \frac{\pi}{4} d_p^2 (N_1 + \Delta N_1) + \bar{k}_2 \frac{\pi}{4} \bar{d}_{2p}^2 \Delta N_2. \tag{14}$$

Substituting Eqs. (12) and (13) into Eq. (14), and rearranging,

$$-\log T = k_1 \frac{\pi}{4} d_p^2 \left(N_1 - K_{11} N_1^2 \Delta t + \frac{1}{2} \alpha K_{11} N_1^2 \Delta t \right). \tag{15}$$

Then,

$$\Delta(-\log T) = (-\log T) - (-\log T_1)$$

$$= -k_1 \frac{\pi}{4} d_p^2 K_{11} N_1^2 \left(1 - \frac{\alpha}{2}\right) \Delta t, \qquad (16)$$

$$\Delta(-\log T/ -\log T_1) = -K_{11} N_1 \left(1 - \frac{\alpha}{2}\right) \Delta t$$

$$\frac{d(-\log T/ -\log T_1)}{dt}\bigg|_{t=0} = -K_{11}\left(1 - \frac{\alpha}{2}\right) N_1. \qquad (17)$$

Fig. 6 shows an example of the rate of flocculation of the *Chlorella* cells in the agitated vessel (Fig. 1). By drawing tangents at the origin to each curve in the figure and by determining the slope, the values of the reaction-rate at $t = 0$ can be plotted against N_1 in a logarithmic paper (see solid circles in Fig. 7). The flocculation-rate data which are obtained with the same rotation speed, $n = 277$ rpm of the impeller, different in the metallic ion condition from Fig. 6 are rearranged similarly as are also plotted in Fig. 7.

It seems acceptable to draw the straight lines in Fig. 7 to correlate the initial rate of the flocculation with the initial number concentration, N_1 of single cells. The fact that each slope of the solid line in the figure is nearly equal to unity justifies Eq. (17) – the flocculation is of the second order reaction –.

The value of $K_{11}\left(1 - \frac{\alpha}{2}\right)$ in the right-hand side of Eq. (17) is assumed definite, if the condition of a specific flocculation experiment be given. Yoshizawa calculated the ratio of projected area with respect to a projection of flocs comprising two spherical particles onto the single spheres. According to his calculation, the value of α is estimated as [7]:

$$\alpha = 1.85.$$

If this value of α is substituted into Eq. (17),

$$-\frac{d(-\log T/ -\log T_1)}{dt}\bigg|_{t=0} = (0.075) K_{11} N_1. \qquad (18)$$

It is, now, possible to assess from the flocculation-rate data the reaction-rate constant, K_{11} as affected by the various operation conditions (rotation speed of impeller, ion conditions, pH value and so forth) (cf. Fig. 7 and Eq. (18)).

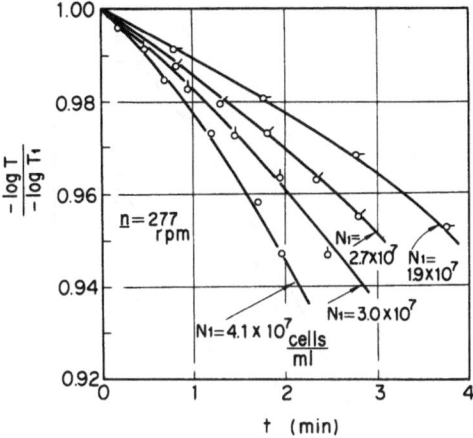

Fig. 6. Rate of flocculation (*Chlorella* cells)

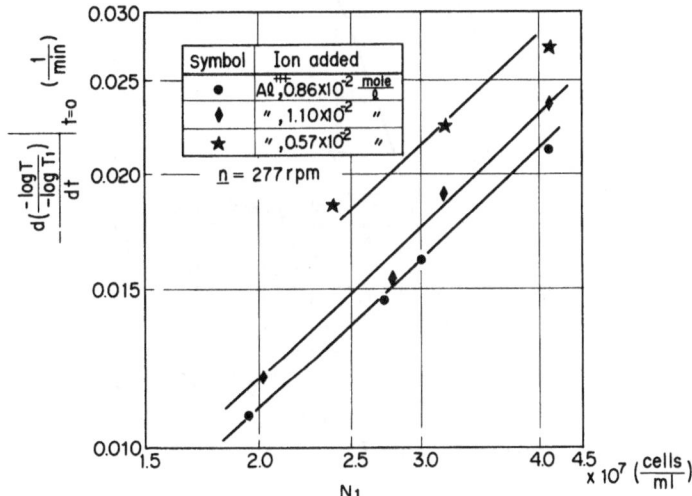

Fig. 7. Rate of flocculation (*Chlorella* cells) at $t = 0$

3.2. Effect of Rotation Speed of Impeller on the Reaction-Rate Constant [5, 21]

The effect of rotation speed, n of the impeller (see Fig. 1) on the value of K_{11} for the *Chlorella* cells is shown in Fig. 8. The initial number concentration, N_1 of the cells is adjusted to $N_1 = 4.1 \times 10^7$ cells/ml ($-\log T = 0.5$). The initial pH value was about 4.2, and the value did not show any appreciable change during the period of each measurement (less than 5 min) at room temperature (20° C).

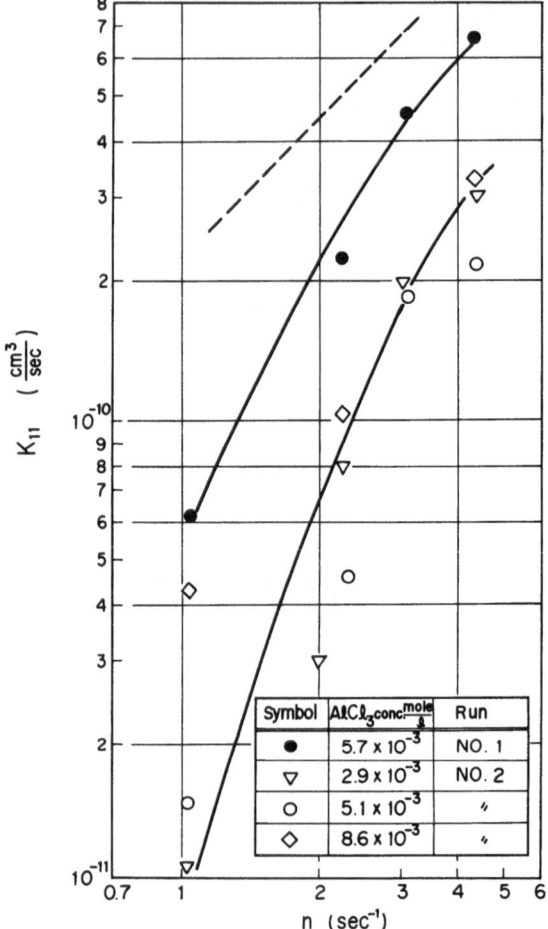

Fig. 8. Effect of rotation speed, n on K_{11}

Though the cultivation procedure of the cells is exactly the same, the values of K_{11} depend appreciably due to unknown reasons on the run. The fact that difference in cultivation-lot far overweighs that of the metallic ion concentration seems typical of the difficulty in reproducing the flocculation-rate data on the viable material.

If the solid curves assumed in Fig. 8 represent the correlation between K_{11} and n, the slope is apparently larger than unity (a broken line whose slope = 1 is shown in the figure for reference).

If the intra-particle collision presented by Smoluchowski [16] who dealt with the subject in a laminar flow region is acceptable, the reaction-rate constant, K_{11} must be:

$$K_{11} = \frac{32}{3} \left(\frac{d_p}{2} \right)^3 \frac{du}{dr} . \tag{19}$$

If the shear rate, du/dr is assumed proportional to the rotation speed, n of the impeller,

$$K_{11} \propto n . \tag{20}$$

Indeed, judging from the modified Reynolds number, N_{Re} which extends from 10^3 to 10^4 regarding the experimental condition for the data points in Fig. 8, the liquid flow in the agitated vessel (Fig. 1) is in a transient region between laminar and turbulent.

An apparent discrepancy of the data points (Fig. 8) from Eq. (20) seems to suggest the necessity of taking the energy barrier $- \zeta$-potential $-$ duly into account in a proper interpretation of the reaction-rate constant. In connection with this discussion, the initial value of pH (of the suspension) which seems to affect remarkably the value of K_{11} must be mentioned.

Besides aluminum and calcium chlorides, sodium hydroxide or hydrochloric acid was used to change the initial pH value of the *Chlorella* suspension, while the ζ-potential of the cells was measured with an electrophoretic cell under a microscope [3, 5]. In spite of the less susceptibility of the ζ-potential to initial pH value of the suspension, the value of K_{11} in an alkaline side increased to an extent of several times as much as in an acidic condition (not shown).

This intriguing observation seems to suggest the significance of the polyhydric ions which result presumably from the various pretreatment of the cells and which are due to be the "bridging" factor in the flocculation. This "bridging" concept in discussing the structure of a microbial floc is in conformity with the previous discussion on water content of the floc. The sophisticated structure of the microbial surface which should not be overlooked makes a consistent and quantitative discussion on the reaction-rate constant, K_{11} exceedingly difficult, indeed.

3.3. Deflocculation of *Chlorella* Cells vs. Shear Rate [21]

Suppose that the flocs of *Chlorella* be exposed to a shear stress due to the liquid viscosity in the agitated vessel (see Fig. 1), keeping a specific value of $\bar{d}_f$ unchanged in the steady state.

A force balance between the shear stress, τ and the pseudo-surface tension, σ_f introduced here to define the value of $\bar{d}_f$ will be:

$$\tau \left(= \mu \frac{du}{dr} \right) = \frac{\sigma_f}{\bar{d}_f} . \tag{21}$$

The implication of the term, σ_f is as follows: Assuming a rectangular parallelopiped whose length and cross-sectional area being unity and d_p^2, the number of single cells in this particular floc is:

$$\frac{d_p^2(1-\varepsilon)}{\frac{\pi}{6}d_p^3}.$$

Designating the number of particle-particle contact per single cells and the adhesive force at each contact to v and F, respectively, the pseudo-surface tension, σ_f is introduced by the following equation.

$$\sigma_f = \frac{d_p^2(1-\varepsilon)}{\frac{\pi}{6}d_p^3}vF = k_3(1-\varepsilon)\frac{F}{d_p}. \tag{22}$$

The shear rate, du/dr is assumed as follows:

$$\frac{du}{dr} = k_4 n. \tag{23}$$

Substituting Eqs. (21) and (22) into Eq. (23), and rearranging,

$$\frac{\bar{d}_f/d_p}{(1-\varepsilon)} = \frac{k_3 F}{\mu k_4 d_p^2}n^{-1} = k_5 n^{-1}. \tag{24}$$

The value of k_5 is assumed constant.

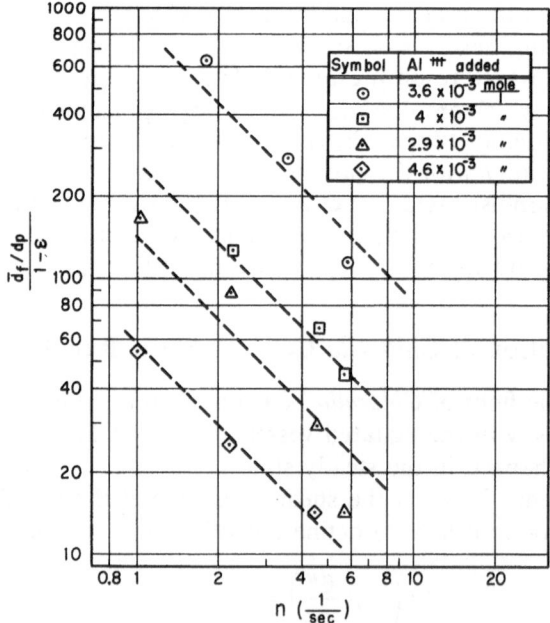

Fig. 9. Floc size as affected by rotation speed of impeller

The experimental data on the floc size, $\bar{d}_f$ as affected by the rotation speed, n of the impeller are shown in Fig. 9. The values of $\bar{d}_f$ in this experiment are determined by converting the values of V_2 millivolt in each steady state into those of turbidity, $-\log T_2$ to use the relationship in Fig. 4 between $-\log T_1/-\log T_2$ and $\bar{d}_f/d_p$ (note: $-\log T_1 = 0.5$, $V_1 = 15$ millivolt and $d_p = 3.2$ microns). Values of $(1 - \varepsilon)$ appearing on the ordinate of Fig. 9 are estimated from the solid curve in Fig. 5.

It is noted from Fig. 9 that the data points in each run are situated along a broken line whose slope being -1 in the figure (cf. Eq. (24)). However, the data points in the figure indicate apparently a more appreciable dependence on n than that shown by Eq. (24). Presumably this fact is attributable to that the shear stress disrupting the flocs is controlled by the liquid turbulence besides the shear due to the liquid viscosity.

4. Sedimentation

4.1. Equivalent Size of Activated-Sludge Floc

Fig. 10 shows some typical settling-patterns of the activated sludge. The pH value of mixed liquor was nearly around 7.0. Except the bulking sludge, the activated sludge used was transferred immediately from

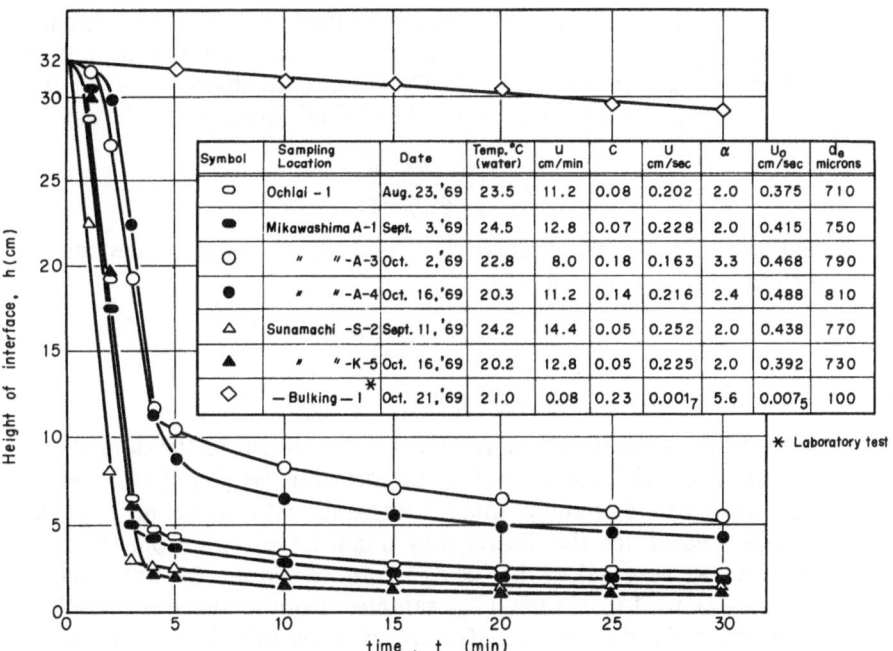

Symbol	Sampling Location	Date	Temp.°C (water)	U cm/min	c	U cm/sec	α	Uo cm/sec	de microns
○	Ochiai – 1	Aug. 23, '69	23.5	11.2	0.08	0.202	2.0	0.375	710
●	Mikawashima A-1	Sept. 3, '69	24.5	12.8	0.07	0.228	2.0	0.415	750
○	" –A-3	Oct. 2, '69	22.8	8.0	0.18	0.163	3.3	0.468	790
●	" –A-4	Oct. 16, '69	20.3	11.2	0.14	0.216	2.4	0.488	810
△	Sunamachi –S-2	Sept. 11, '69	24.2	14.4	0.05	0.252	2.0	0.438	770
▲	" –K-5	Oct. 16, '69	20.2	12.8	0.05	0.225	2.0	0.392	730
◇	— Bulking — I*	Oct. 21, '69	21.0	0.08	0.23	0.001_7	5.6	0.007_5	100

* Laboratory test

Fig. 10. Typical settling data of activated sludge under gravity

each exit of an aeration basin to a measuring cylinder (6.3 cm in internal diameter and 32 cm in graduated height) to observe the rate of interfacial subsidence. The bulking sludge was prepared artificially and observed in a test room.

Assuming straight lines through the data points which were recorded during 1 to 3 minutes after the initiation of settling run (see Fig. 10), the constant rate, u cm/min could be estimated. The constant settling continued nearly for a whole range of observation (30 min) in the bulking sludge. For this particular sludge the height of a sediment in the cylinder after 24 hr was defined to give the fraction, c of the sludge, while for other sludges the sediment height, h at $t = 30$ min was taken to define the value of c, because no appreciable change of h was observed afterwards [18].

The hindered settling velocity, U cm/sec of sludge was estimated.

$$U = \frac{u}{60(1-c)} \,. \tag{25}$$

Taking the effect of interference of neighbouring flocs on the value of U into account, the free settling velocity, U_0 of single flocs was assessed by the following equation [2].

$$U_0 = U(1 + \alpha c^{1/3}) \,. \tag{26}$$

An empirical factor, α manifesting the effect of mutual interference on the settling velocity of flocs was estimated from the correlation ever published between α and c [1, 2]. The equivalent diameter, d_e of the sludge floc could be calculated from the Stokes' formula as follows:

$$d_e = \sqrt{\frac{18 U_0 \mu}{g(\varrho_y - \varrho)}} \,. \tag{27}$$

Taking 22° C as the average temperature of liquid of the mixed liquor (see the data inset in Fig. 10),

$$\mu = 9.5 \times 10^{-3} \text{ g/cm sec} \,,$$
$$\varrho_y = 1.003 \text{ g/cm}^3 \text{ [2]} \,,$$
$$\varrho = 1.000 \text{ g/cm}^3$$

were taken to calculate the value of d_e as also shown in the data sheet.

Values of d_e which appeared in the previous work [2] ($d_e = 60$ to 100 microns) are extremely small compared with those in Fig. 10. A principal reason for the discrepancy might have come from that the previous data were taken by using a mixed liquor (return sludge, MLSS = 5,200 to 15,500 ppm [2]) sampled from a sewage-treatment plant, whereas the data in Fig. 10 dealt with the mixed liquor prior to the secondary sedimentation basin. In fact, the value of MLSS in the current example ranged from 780 to 1,750 ppm (cf. Table 2 later). The concentration process in terms of return sludge might have mainly been responsible for minimizing the activated sludge flocs.

Except the bulking sludge, no distinct difference in the settling patterns is noted at first glance from Fig. 10. However, the point which remains to be discussed here is that the quality of water, i.e., the supernatant liquid in the settling operation differs appreciably depending on the ecological type of the sludge from case to case as will be elaborated below.

4.2. Species of Fauna (Flora) of Floc vs. Effluent Quality [18]

The removal of bacteria regarding the typical sludge-settling data in the previous figure is shown in Table 2. The value of transparency, κ cm was determined by transferring each effluent of the secondary sedimentation basin to another graduated and standardized cylinder [17]. Clearly, the supernatant liquid from the bulking sludge is most transparent, though the bulking phenomenon is prohibiting in the plant operation principally due to the extremely slow settling rate of microbial flocs from the mixed culture.

Effluents from Ochiai, Mikawashima A-1 and A-3 sludges are fairly good, while those from other plants are least recommendable due to the lower values of κ as shown in the table. In fact, the larger value of κ is accompanied by the lower values of BOD or COD, although the latter criteria are omitted from the table.

The above argument must be discussed further from the viewpoint of bacterial removal [12]. The 5th and 6th columns in Table 2 are the values of η_1 and η_2 for the bacterial removal efficiencies with respect to the sludge settling operations.

It is interesting to remark from the table that the larger values of κ guaranteed the excellent efficiencies, as a whole, of the bacteral removal in settling and *vice versa*. To make further insight to a relationship between the values of κ and η_1 (or η_2) in Table 2, the microbes comprising each sludge floc were examined especially from the standpoint of fauna. A sample (0.05 ml) was transferred to a slide glass (with parallel streaks, 1 mm in interval) for a microscopic observation.

Dominant genera of fauna and flora observed so far are shown in the 7th column in Table 2. Fig. 11 is a schematic representation of the sludge floc, characterized by the dominant species of fauna [18].

Circles in the figure simulate the sludge flocs, the diameters of which are taken as 700 and 200 microns for the ordinary and bulking sludges,

Table 2. Quality of effluent in the settling of activated sludge

Sample	MLSS mg/l	d_e microns	κ^a cm	η_1^b	η_2^c	Dominant microbes in microscopic obs.	Ecological type and No.^{d}
Ochiai-1	870	710	75	0.99	0.95	Opercularia, Epistylis	Fixing (Colonial), No. 1
Mikawashima-A-1	780	750	86	1.00	0.99	Carchesium, Epistylis	Fixing (Colonial), No. 1
Mikawashima-A-3	1,190	790	55	0.99	0.97	Vorticella	Fixing (Solitary), No. 2
Mikawashima-A-4	920	810	19	0.97	0.85	Amoeba	Crawling, No. 5
Sunamachi-S-2	1,350	770	26	0.97	0.71	Uronema, Tetrahymena	Free swimming (Ciliata), No. 4
Sunamachi-K-5	1,160	730	16	0.97	0.72	Bodo, Oikomonas	Free swimming (Flagellata), No. 3
– Bulking-1	1,750	100	more than 100	1.00	more than 0.99	Sphaerotilus, Fungi	Filamentous, No. 6

a Transparency of effluent measured in situ with the standard method [17] except for bulking sludge.

b $\eta_1 = 1 - \left(\dfrac{\text{Number concentration of viable bacteria in supernatant liquid after settling under gravity}}{\text{Number concentration of viable bacteria in mixed liquor}} \right)$.

c $\eta_2 = 1 - \left(\dfrac{\text{Number concentration of coliform bacteria in supernatant liquid after settling under gravity}}{\text{Number concentration of coliform bacteria in mixed liquor}} \right)$.

d See Fig. 11.

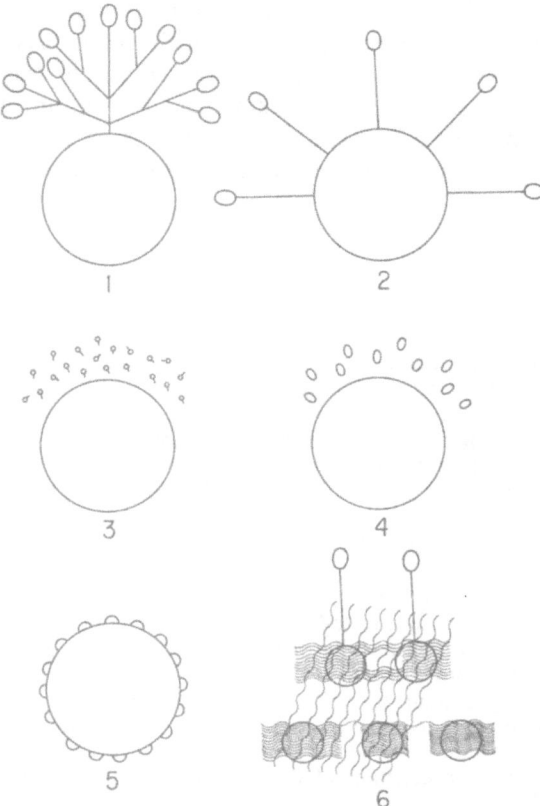

Fig. 11. Schematic representation of sludge floc with particular reference to fauna

respectively with reference to quite a few estimations other than those exemplified in Table 2. The size and schematic shape of each protozoa attached to and/or swimming around the floc are also drawn in proportion to their actual pictures taken by the photomicrographs.

The last column of Table 2 designates the ecological type which was advocated by Sudo et al. [18] referring to each dominant genus in fauna observed in the sludge examination. Indeed, the fixing type, colonial in particular exhibited the most recommendable characteristics of both the settling pattern (Fig. 11) and the quality of effluent (Table 2).

The reason for this good performance of the sludge could be attributed to the ramification of the floc surface due to the specific protozoa, whereas the poor performance of the swimming type especially in terms of the bacterial removal in settling might be ascribed to a less possibility of entrapping the bacteria due to the free swimming

4*

characteristics of protozoa. The excellent removal of the bacteria by the bulking sludge can be expected from the same reasoning.

The most exciting problem to be dissolved in the future in this specific separation of microbial cells from the culture medium in this category seems to be in finding out a clue, if any, to control the emergence of the desirable microbial species of fauna.

Symbols

c	volume fraction of cellular aggregate
D_i	impeller diameter, cm
d_e	equivalent diameter of floc, microns
$\bar{d}_f$	mean surface diameter of floc, microns
$d_{f \cdot n}$	nominal diameter of activated sludge floc, microns
$d_{f \cdot v}$	mean volume diameter of floc, microns
d_p	average diameter of microbial (*Chlorella*) cell or particulate particle, microns
$\bar{d}_{2p}$	mean surface diameter of floc composed of two pieces of cells, microns
F	adhesive force between particles, dyne
g	acceleration due to gravity, 9.80 m/sec^2
h	height of interface in test cylinder, cm
I	intensity of light passing through cell suspension, measured with photo-electric photometer
I_0	intensity of light through cell-free water as control
K_{11}	reaction-rate constant of flocculation between single cells, initial reaction-rate constant, cm^3/sec
K_{ij}	reaction-rate constant of flocculation which concerns with N_i and N_j, cm^3/sec
K_{kj}	reaction-rate constant of flocculation which deals with N_k and N_j, cm^3/sec
k_1	proportionality constant
k_2	proportionality constant
$\bar{k}_2$	proportionality constant
k_3	$6v/\pi$, empirical coefficient
k_4	proportionality constant
k_5	$k_3 F/\mu k_4 d_p^2$, coefficient
m	empirical exponent
N	number concentration of single cells or floc, $1/\text{cm}^3$
N_1	number concentration of single cells, initial number concentration of single cells in flocculation, $1/\text{cm}^3$
N_2	number concentration of flocs, final number concentration of flocs in flocculation, $1/\text{cm}^3$
N_i	number concentration of flocs; each floc is constructed by i pieces of cells, $1/\text{cm}^3$
N_j	number concentration of flocs; each floc comprises j pieces of cells, $1/\text{cm}^3$
N_k	number concentration of flocs; each floc is composed of k cells, $1/\text{cm}^3$
N_{Re}	$n D_i^2 \varrho / \mu$, modified Reynolds number
n	rotation speed of impeller, $1/\text{sec}$
$\underline{n}$	rotation speed of impeller, $1/\text{min}$
R	electric resistance of CdS, kiloohm

R_1 electric resistance of CdS for $-\log T_1$, kiloohm
R_2 electric resistance of CdS for $-\log T_2$, kiloohm
$-\log T$ $\log I_0/I$, turbidity
$-\log T_1$ $\log I_0/I_1$, initial turbidity of cell suspension $(=0.5)$
$-\log T_2$ $\log I_0/I_2$, final turbidity of cell suspension
t time, sec, min
U settling velocity of swarm of particles under gravity, cm/sec
U_0 settling velocity of single particles under gravity, cm/sec
u rate of interfacial subsidence under gravity, cm/min
du/dr velocity gradient of liquid medium, 1/sec
V recorder reading, millivolt
V_1 initial recorder reading, millivolt
V_2 final recorder reading, millivolt
X cell mass concentration, mg/cm^3

Greek letters

α $\bar{k}_2 \bar{d}_{2\,p}^2 / k_1 d_p^2$, constant; correction factor (hindered settling)
δ characteristic value of CdS
ε water content of floc in fraction
η_1 collection efficiency for viable bacteria
η_2 collection efficiency for coliform bacteria
κ transparency, cm
μ viscosity of continuous phase of suspension, viscosity of liquid (water), g/cm sec
ν number of particle-particle contact per single cells
ϱ liquid density, g/cm^3
ϱ_p cell density, g/cm^3
ϱ_s true density of solid, g/cm^3
ϱ_y floc density, g/cm^3
σ_f pseudo-surface tension, dyne/cm
τ shear stress, dyne/cm^2

References

1. Aiba, S., Kitai, S., Ishida, N.: J. Gen. Appl. Microbiol. **8**, 109 (1962).
2. — — Heima, N.: J. Gen. Appl. Microbiol. **10**, 243 (1964).
3. — Someya, J.: J. Ferm. Technol. **46**, 387 (1968).
4. — Watanabe, T., Hirata, M.: J. Ferm. Technol. **48**, 125 (1970).
5. Hirata, M.: Master's Thesis, Dept. Chem. Eng., Univ. of Tokyo (1969).
6. Japan Patent S. 42-27309 (1967).
7. Mori, Y., Yoshizawa, A.: Chem. Eng., Japan **31**, 1132 (1967).
8. Mueller, J. A., Voelkel, K. G., Boyle, W. C.: J. Sanitary Eng. Division, Proc. ASCE **92**, SA 2 (1966).
9. Nagai, S., Igarashi, M., Taguchi, H., Teramoto, S.: J. Ferm. Technol. **41**, 1 (1963).
10. Nagatani, M., Kuba, Y.: Presented at the Regional Meeting, Soc. Ferm. Technology, June (Tokyo) (1969).
11. — — Takamiya, Y., Sugama, S.: J. Soc. Brewing, Japan **65**, 344 (1970).

12. Prakasam, T. B. S., Dondero, N. C.: Appl. Microbiol. **15**, 461 (1967).
13. Reich, I., Vold, R. D.: J. Phys. Chem. **63**, 1497 (1959).
14. Shirato, S., Esumi, S.: J. Ferm. Technol. **41**, 87 (1963).
15. — Suzuki, I., Esumi, S.: J. Ferm. Technol. **43**, 501 (1965).
16. Smoluchowski, M. V.: Z. Physik. Chem. **92**, 129 (1917).
17. Standard Methods for the Examination of Sewage, p. 33. Tokyo: Japan Sewage Works Ass. (1964).
18. Sudo, R., Aiba, S.: J. Ferm. Technol. **48**, 342 (1970).
19. Tambo, N., Watanabe, Y.: J. Water Works Ass. (Japan), No. 397, 2 (1967).
20. Thomas, D. G.: A.I.Ch.E. Journal **9**, 310 (1963).
21. Watanabe, T.: Master's Thesis, Dept. Chem. Eng., Univ. of Tokyo (1968).
22. Yoshizawa, A.: Chem. Eng., Japan **32**, 1233 (1968).

S. Aiba and M. Nagatani
Institute of Applied Microbiology,
University of Tokyo, Tokyo, Japan
Institute of Brewing,
Ministry of Finance, Tokyo, Japan

CHAPTER 3

A Simplified Kinetic Approach to Cellulose-Cellulase System

T. K. GHOSE and K. DAS

With 7 Figures

Contents

1. Introduction

Perhaps one of the least known areas of enzyme kinetics relates to insoluble substrates. Little literature information is available on the nature of heterogeniety of the substrates and the confirmed mechanism involved. There are three groups of insoluble substrates, namely, steroids, starch, and cellulose. A model for the kinetics of steroid conversion has been proposed (Chen *et al.*, 1962); the mechanism and mode of action of starch degrading and synthesizing enzymes have been discussed (Greenwood, 1968); and there is information on the hydrolysis of starch to maltose (Dawes, 1962). Published data (Alberty, 1956; Wong and Hans, 1962; Whitaker, 1963) on cellulose hydrolysis are speculative because the biochemical machanism is obscure. According to one report (Flora, 1964), the kinetic pattern of cellulose hydrolysis appears to be independent of product inhibition and of the presence of inhibitors in the crude enzyme. Others (Dixon and Webb, 1958) reported similar observations. According to them a linear relationship exists between the degree of hydrolysis and the square root of the contact time between the enzyme and substrates, or

$$S \propto \mu \sqrt{t}.$$

Where S = percent of substrate-hydrolysed,
  t = contact time,
  μ = rate of hydrolysis (a function of the nature of cellulose).

It has also been shown that the rate (μ) decreases with increasing crystallinity of the cellulose, i.e., amorphous (swelled) cellulose gives a larger rate than hydro-cellulose or cotton. A model for cellulose hydrolysis (Amemura and Terui, 1965) suggests that the rate of overall substrate dissolution is limited neither by adsorption rate nor by the penetration of enzyme molecule. It is also claimed that there exists a linear function between the crystalline substrate surface destroyed and the square of the surface area available for contact (King, 1966).

Accordingly,

$$\frac{dA}{dt} = \alpha(S_0)^2$$

where, A = substrate surface destroyed,
  S_0 = substrate surface available,
  t = time of contact,
  α = rate constant for surface destruction.

This suggests that the enzyme action is not exclusively a surface phenomenon because the total number of particles per unit volume of reaction system should have remained constant until particles too small to be detected were produced. King, however, reported that there is a marked increase in the total number of particles during the initial stages of reaction. This increase is attributed to a fragmentation of the larger initial particles into 800–1500 small ones. This most probably is due to the action of C_1 component of the enzyme on the cellulose microfibrils. However, the exact number of fragmented particles could not be determined. Since both the surface erosion and fragmentation action appear to occur at the same time the phenomenon cannot be considered as due to fragmentation alone (Halliwell, 1965).

It has been reported (Stone et al., 1969) that a cellulase molecule has a diameter of 30 Å. When the substrate pore volume accessible to the enzyme is zero, the reaction rate is also zero, but, for any additional increase in the accessible pore volume, there is a corresponding increase in reaction rate. Fig. 1 shows that molecules of sizes smaller than 30 Å produce positive intercepts on the Y-axis (accessible pore volume, ml/g, against initial rates, % dissolved per hour) because more water is

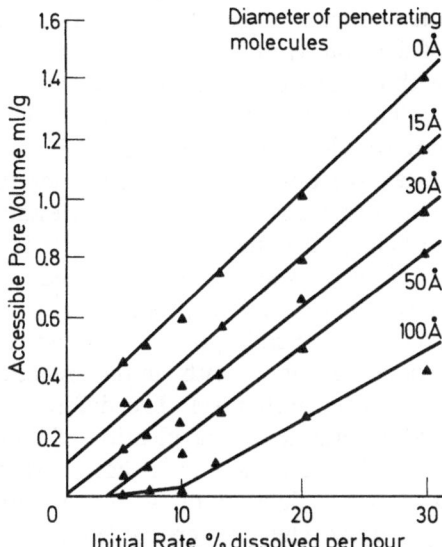

Fig. 1. Relationship between the initial rate of reaction and the volume of water accessible to molecules of different sizes

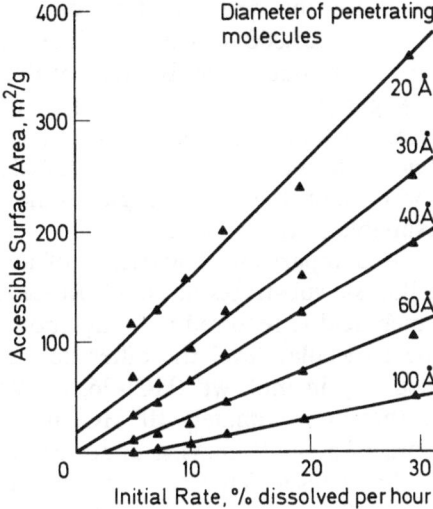

Fig. 2. Relationship between the initial rate of reaction and the surface area of cellulose accessible to molecules of different sizes

accessible to the polymer than to the enzyme molecule. Molecules larger than 30 Å produce negative intercepts. According to these authors, similar relationships appear to exist in terms of accessible surface area (m²/g) as a function of initial rates (Fig. 2). It appears from the two approaches that the digestibility of various celluloses of different degrees of swelling (in terms of volume and area) is directly proportional to the accessibility of molecules of 30–40 Å in diameter which might well be the diameter range of cellulase molecules. Reports (Whitaker *et al.*, 1954) of data on *Myrothecium verrucaria* cellulase appear to be in agreement with the observations of Stone *et al.*

The nature of native cellulose, whether crystalline or amorphous, is not yet clear. It is reported (Mikhailov, 1958) from an examination of X-ray patterns of native celluloses that, unlike regenerated celluloses whose phase structures are well established, there is no definite reason to believe that the native polymer is crystalline in nature. Since initial reaction rates are primarily related to the physical state of the cellulose, it remains an open question as to how cellulase works on cellulose. Neither accessible pore volume nor surface (by virtue of their demonstrated linearity with initial reaction rates) can fully explain the intriguing mechanism. The difference between the mechanism of acid and enzymatic saccharification of cellulose has been reported (Walseth, 1952). It has been shown that enzymatic hydrolysis of cellulose causes a slower reduction of degree of polymerization (DP) than the acid hydrolysis. The most probable reason for such a difference is the relative size of the two catalysts (enzyme and mineral acid) and their ability to permeate the fine structure of cellulose. Large cellulase molecules (mol. wt. 25,000–67,000) are likely to penetrate only into the larger intercrystalline space. Because of their high catalytic activity, all chain linkages (β-1, 4 glucosidic) which can be contacted are readily hydrolyzed to form soluble sugars. On the other hand, the relatively small acid molecules are capable of diffusing into the smaller intercrystalline spaces and then attack glucosidic bonds which are otherwise not available to the enzymes.

A comparison of activities of acid and enzyme catalysts on three cellulosic substrates at 50° C (Reese, 1956) shows that 100,000 times as much acid is required to bring about the same degree of hydrolysis. At the molecular level the difference is further increased because of the disparity in mol. wt. (HCl-36, cellulase 63,000; Whitaker *et al.*, 1954) so that approximately 10^8 HCl molecules are required to perform the work of a single enzyme under certain given conditions.

Enzymes which catalyze the hydrolysis of simple soluble compounds may split a million (or more) bonds per minute per enzyme molecule. Molecules in solution pose no problem of accessibility to enzyme. What happens when the substrate molecule gets much larger, i.e., as one moves from the monomer to polymer? The interaction between large enzyme and large substrate molecules may be somewhat impeded, i.e. it is likely to be more difficult for the enzyme to accommodate itself to the site at which reaction is to occur. But even here the rate of catalysis is high, in the neighborhood of 18,000 bonds per minute per enzyme molecule for α-amylase and 240,000 bonds for β-amylase. Viscous polymer and branched polymers present additional impedence to enzyme attack, the first through restricting enzyme movement and the second through blocking hydrolysis sites. Substituents on the substrate may also bind enzyme through electrostatic charges, further reducing the rate of hydrolysis.

2. Product Inhibition

In most microbiological and biochemical systems, accumulation of end products exercises an inhibitory effect on the rate of the forward reaction. One of the major products of hydrolysis of cellulose is cellobiose, which stimulates C_x activity of *Streptomyces* spp. filtrates when the substrate is solubilized by the introduction of various substituents, e.g., CMC, hydroxycellulose, cellulose acetate, etc. Stimulation is absent when unsubstituted cellulose is used (Reese *et al.*, 1952); and product inhibition is common. Cellulase is competitively inhibited by cellobiose and methocel and inactivated by some protein reactants as halogens, heavy metals and detergents (Mandels and Reese, 1965). Cellobiose inhibited the hydrolysis of cellulose by filtrates of most of the 36 organisms tested by these authors. This action of cellobiose is believed to proceed in much the same manner that maltose inhibits hydrolysis of starch by amyloglucosidase through a cleavage at the $\alpha,1$–4 and $\alpha,1$–6 glucosidic linkages. The inhibitory effect of products varies with the organism from which the cellulase is derived. Thus, lactose is a very good inhibitor of the enzyme from *Penicillium pusillum* (Table 1), but cellobiose is a good inhibitor of cellulases of many origins. Glucose inhibition is generally weak. For *Trichoderma viride* cellulases acting on finely milled and heat-treated cellulose, a concentration of 30% glucose gives only 40% inhibition (Ghose, 1969a). Product inhibition of the cellulose-cellulase system is reported (Nisizawa *et al.*, 1963) to be greatest for the larger soluble oligomers (cellopentaose) and diminishes rapidly

Table 1. *Inhibition of Penicillium pusillum cellulase by sugars* (Mandels and Reese, 1963)

Sugar 1%	Inhibition of hydrolysis %			
	CMC visc.	Swollen cellulose weight loss	Ball-milled cotton weight loss	Cotton swelling
Cellobiose	48	46	72	52
Lactose	88[a]	44	60	54
Maltose	5	4	1% stim.	NT
Glucose	5	NT	NT	37

[a] Lactose (0.1%) gave 52% inhibition; NT = Not tested.

with decreasing product molecular sizes. There would be a definite advantage in having cellobiase in the cellulose hydrolysates for a reduced dimer inhibition; this inhibition is, however, much less than that exercised by glucose. Continuous removal of the products from the system without loss of the substrate or the enzyme has been effected by the use of molecular sieve membranes (Ghose, 1969c).

The nature of this inhibition has not been explained. Product inhibition in the case of dehydrogenation of dienedol and of hydrocortisone by cells of *Septomyxa affinis* is also reported (Chen *et al.*, 1962). For product inhibition of cellulose-cellulase system, no linearity is noticeable. Data (Ghose, 1969d) on high rate saccharification of cellulose at lower substrate concentrations are presented in Table 2 and Fig. 3. The plots of $\frac{1}{x}\log\frac{a}{a-x}$ vs. $\frac{t}{x}$ (Fig. 4) for cellulosic substrate of 5%, 2% and 1% consistencies in concentrated cellulase (Tv) does not give the expected linearity.

Table 2. *Enzymatic saccharification of fine cellulose in concentrated Trichoderma viride cellulase*

Initial substrate concentration a, g/l	Reaction time t, min	Substrate concentration reacted x, g/l	Substrate concentration un-reacted $(a-x)$, g/l	$\log\dfrac{a}{a-x}$	$\dfrac{1}{x}\log\dfrac{a}{a-x}$	t/x
10	0	10.0	0	—	—	—
	30	6.86	3.14	0.503	73.0	4.37
	135	5.09	4.91	0.309	60.7	26.85
	315	3.55	6.45	0.190	53.6	88.74
	585	2.48	7.52	0.124	50.0	235.8
	2430	1.31	8.69	0.061	46.6	1855
	3165	1.05	8.95	0.048	45.7	3014
20	0	20	0	—	—	—
	30	14.86	5.14	0.590	40.6	2.02
	135	11.96	8.04	0.396	33.1	11.30
	315	9.48	10.52	0.279	29.42	33.23
	585	7.44	12.56	0.202	27.19	78.62
	1305	5.90	14.10	0.152	26.60	221.2
50	0	50	0	—	—	—
	30	38.90	11.10	0.654	16.81	0.771
	135	32.15	17.85	0.447	13.92	4.26
	315	27.30	22.70	0.343	12.56	11.53
	585	23.80	26.20	0.281	11.79	24.58
	1305	17.95	32.05	0.193	10.75	72.68
	2430	16.60	33.40	0.175	10.55	146.4
	3165	14.40	35.60	0.148	10.24	221.8

Note: Substrate particle size $= -4.7\,\mu$; cellulase activity 380 cx units/ml; Reaction vol. 10 ml containing 0.01% merthiolate, temp. 50° C.

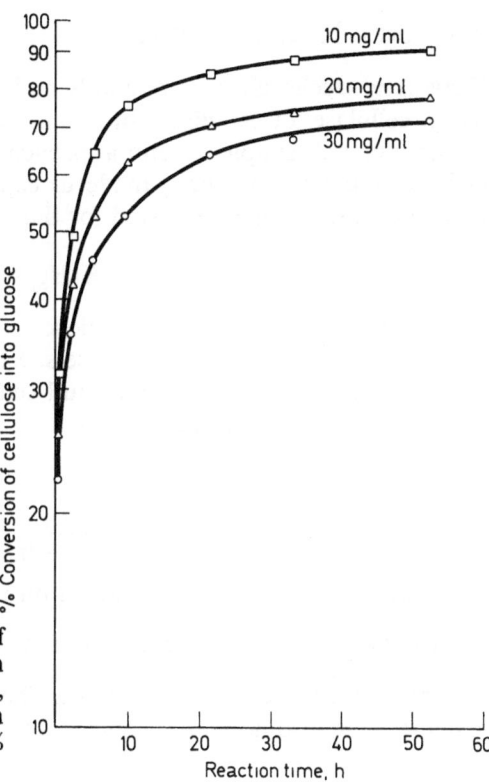

Fig. 3. Product inhibition of cellulose-cellulase system with <4.7 μ cellulose (Solka Floc), activity 380 c_x units/ml, reaction vol. 10 ml containing 0.01% merthiolate, temp. 50° C

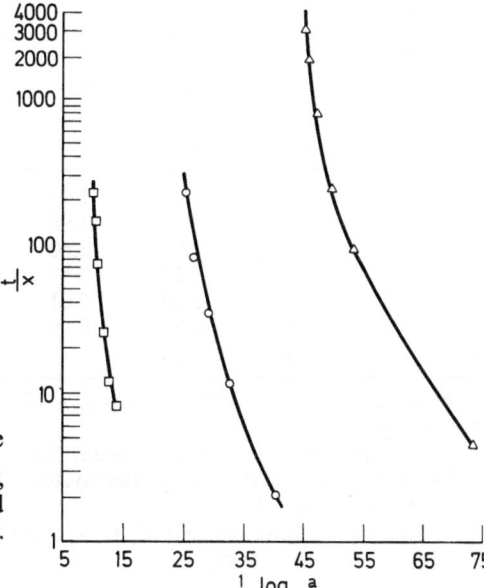

Fig. 4. $\frac{t}{x}$ Vs $\frac{1}{x} \log \frac{a}{a-x}$ plot. (see Table 2). –□–□–□– –5% cellulose, –◯–◯–◯– –2% cellulose and –△–△–△– –1% cellulose in conc. Tv cellulase

3. Overall Reaction Order

Enzymatic saccharification of paddy hulls has been reported (Das, 1969) using cellulase of *A. niger* and a mixed cellulase of *A. niger* and *C. globosum*. A simple kinetic approach to derive rate constants and order of reactions has been made to explain the system. A first order kinetic equation has been tested for the same. Accordingly:

$$\frac{dH}{dt} = K(H_{co} - H_c). \tag{1}$$

Where, H_{co} = initial conc. of cellulose,
  H_c = conc. of hull cellulose saccharified,
 $H_{co} - H_c$ = prevailing conc. of hull cellulose,
  t = reaction time,
  k = rate constant.

Integrating (1), we get

$$Kt = \ln \frac{H_{co}}{H_{co} - H_c} \tag{2}$$

Data presented in Table 3 are based on information drawn from three experiments (details described). Plots of $\log \dfrac{H_{co}}{H_{co} - H_c}$ Vs. t are linear (Fig. 5) for the three different systems. The values of k for the three cases a, b and c are also noted on Table 3. The difference in the values of the

Table 3. *Enzymatic saccharification of milled rice (Das, 1969) hulls in cellulase from various sources*

Reaction time, t hrs.	Glucose produced mg/10 mg hull cellulose			% Conversion of hull cellulose into sugar		
	a	b	c	a	b	c
3	0.49	0.49	1.07	4.45	4.45	9.73
6	1.19	0.97	1.98	10.81	8.82	18.00
12	2.06	1.85	3.60	18.72	16.81	32.73
24	4.06	3.22	5.57	36.90	29.30	50.63
40	5.60	5.10	7.01	50.90	46.36	63.73
60	6.91	6.72	8.49	62.81	61.00	77.30
70	7.00	6.83	—	63.61	62.00	—

Note: Paddy hulls (40.5% cellulose, $-175\,\mu$) saccharified in cellulase in 5% consistency, pH 4.2.
  a = *A. niger* cellulase, agitated system; Temp. $-25°$ C ($K = 2.99 \times 10^{-4}\,\text{min}^{-1}$).
  b = *A. niger* cellulase, non-agitated system; Temp. $-25°$ C ($K = 2.49 \times 10^{-4}\,\text{min}^{-1}$).
  c = Mixed cellulase (50:50) of *A. niger* and *C. globosum*, agitated system at $30°$ C ($K = 5.53 \times 10^{-4}\,\text{min}^{-1}$).

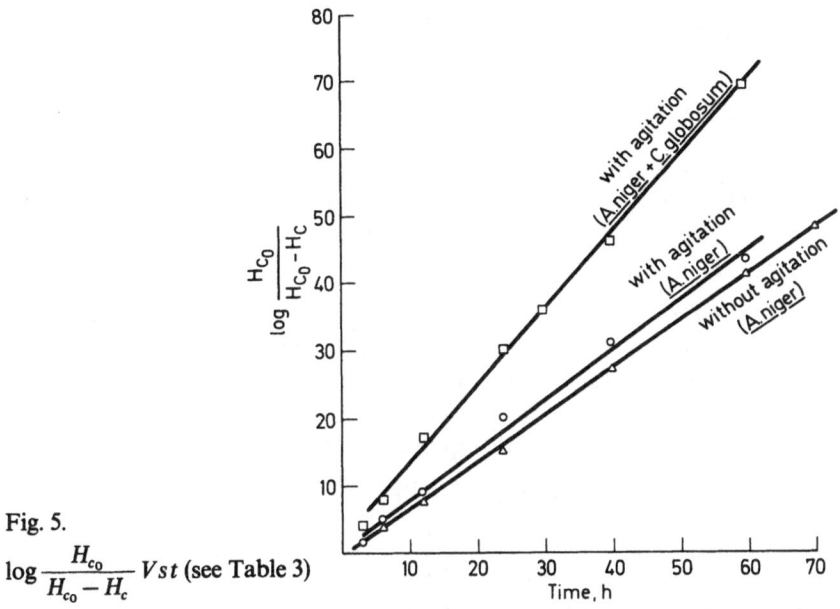

Fig. 5.

$\log \dfrac{H_{c_0}}{H_{c_0} - H_c}$ Vst (see Table 3)

rate constants between *A. niger* cellulase and mixed cellulases is due to the enzyme systems rather than the temperature. The mixed enzyme system is likely to provide a better distribution of specific cellulase components for the accelerated kinetics. Because of the lack of both knowledge of C_1 components and an absolute method for their estimation in the total cellulase system, it does not seem possible to indicate much about the effectiveness of the complex.

4. Michaelis-Menten Mechanism Applied to Cellulose-cellulase System

The variations in the initial reaction velocity with substrate concentrations in the case of most enzyme systems are explained by *Michaelis-Menten* kinetics. If the effect of pH, non-substrate components and buffers are neglected, the following simplified form can represent the system:

$$(E) + (S) \underset{K_2}{\overset{K_1}{\rightleftarrows}} (ES) \xrightarrow{K_3} (E) + n(P).$$

Where (E) = cellulase concentration,
(S) = initial substrate concentration,
(ES) = cellulose-cellulase complex concentration,
n = number of moles of product,
(P) = product.

The steady state velocity v, is given by:

$$v = \frac{V}{1 + \dfrac{K}{(S)}},$$ (3)

where $V = K_3(E_0)$
  (E_0) = initial cellulase concentration,
  (S) = initial cellulose concentration,
and K = Michaelis constant.

K is a convenient quantity for summarizing experimental results as it represents the value of substrate concentration at which 50% of the maximum velocity (V_m) under the specified experimental condition is reached. Considering the initial rates in the case where initial product concentration is small, the rate equation may be written as:

$$v = -\frac{d(S)}{dt} = \frac{d(P)}{dt} = K_3(ES).$$ (4)

The equilibrium concentration of the complex (ES) may be obtained from the expression of the dissociation.

$$K = \frac{K_2}{K_1} = \frac{(E)(S)}{(ES)}.$$ (5)

The enzyme balance existing during the reaction may be expressed as:

$$E_t = (E) + (ES).$$ (6)

Where (E_t) and (E) are the total and free enzyme concentrations respectively. Combining (5) and (6) we get:

$$(ES) = \frac{(E_t)(S)}{(S) + K}$$ (7)

or

$$v = K_3(E_t)(S)/(S) + K.$$ (8)

Since $V_m = K_3 E_t$, the reaction rate may be written as:

$$v = V_m \cdot \frac{(S)}{(S) + K}.$$ (9)

Since K represents the overall reaction and because $K_2 \approx K_3$, K may also be expressed in terms of individual rate constants (Briggs and Haldane, 1925).

$$K = \frac{K_3 + K_2}{K_1}.$$ (10)

Combining Eqs. (8) and (10), a simplified expression for reaction rate results:

$$v = \frac{K_3(E_t)(S)}{\dfrac{K_3 + K_2}{K_1} + (S)} = \frac{V(S)}{K + (S)}.$$ (11)

The Eq. (9) gives the expected hyperbolic functions of v versus (S). However, V cannot be accurately determined from the $v - (S)$ plots for the simple reason that the values of V_m are the limits of asymtotes.

The reciprocals of these quantities (v, s) when plotted (Lineweaver-Burk diagram) according to

$$\frac{I}{v} = \frac{K}{V} \frac{I}{(S)} + \frac{I}{V}$$

result in stright lines. Slopes (K/V) of these plots are dependent on the mechanism of the enzyme substrate relationships as well as on the involvement of any inhibitory effects arising out of the substrates, products or some other built-in components of the system.

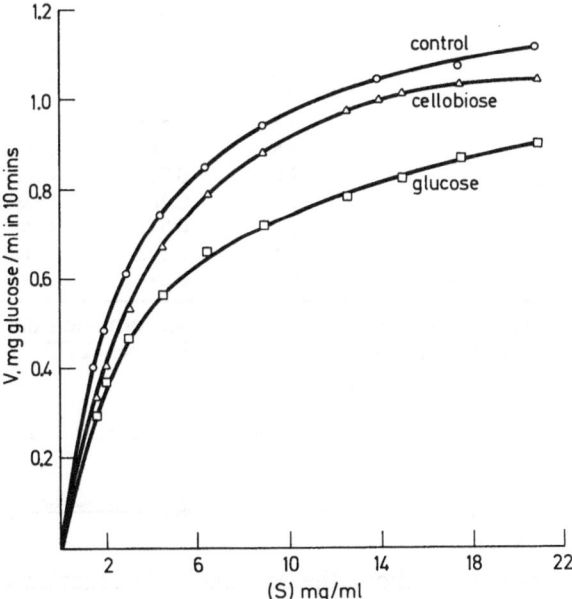

Fig. 6. Glucose and cellobiose inhibition of enzymatic saccharification of cellulose. $<25\,\mu$ cellulose (Solka Floc) in 10 ml tubes, agitated, O-O-O-O-control, $\triangle$-$\triangle$-$\triangle$-$\triangle$-control + 5 mg/ml of cellobiose, $\square$-$\square$-$\square$-$\square$-control + 10 mg/ml of d-glucose. Reaction time -10 min, Temp. $-50°$ C, Cellulase -1.86 FP

Table 4. *Inhibition of enzymatic saccharification in presence of cellobiose and glucose*

Inhibitor	Concentration mg/ml	S_1 mg/ml	v mg/ml	$\dfrac{1}{S_1}$	$\dfrac{1}{v}$
None	0.00	1.5	0.40	0.667	2.5
		2.0	0.48	0.5	2.08
		3.0	0.61	0.33	1.64
		4.5	0.735	0.222	1.37
		6.5	0.845	0.154	1.19
		9.0	0.935	0.111	1.07
		14.0	1.04	0.071	0.96
		17.5	1.06	0.0577	0.94
		21.0	1.11	0.0476	0.90
Cellobiose	5	1.5	0.34	0.667	2.91
		2.0	0.41	0.50	2.44
		3.0	0.53	0.33	1.89
		4.5	0.67	0.222	1.49
		6.5	0.785	0.154	1.27
		9.0	0.88	0.111	1.13
		12.5	0.97	0.080	1.03
		14.0	0.995	0.071	1.00
		15.0	1.01	0.067	0.99
		17.5	1.028	0.058	0.97
		21.0	1.035	0.047	0.96
Glucose	10	1.5	0.29	0.667	3.45
		2.0	0.37	0.50	2.70
		3.0	0.47	0.33	2.13
		4.5	0.565	0.222	1.77
		6.5	0.66	0.154	1.51
		9.0	0.715	0.111	1.40
		12.5	0.78	0.080	1.28
		15	0.825	0.067	1.21
		17.5	0.865	0.058	1.15
		21.0	0.89	0.047	1.12

Note: Rates were measured in terms of product formation during 10 mins. Substrate = $-25\,\mu$ size Sloka Floc (Ghose, 1969d); Enzyme = 2 FP; Temp. = 50° C.

Cellulose-cellulase systems may be written as (Reese *et al.*, 1952).

$$\text{Native cellulose} \xrightarrow{c_1} \text{Hydrated polyanhydroglucose chains} \xrightarrow{C_x} \text{Cellobiose} \xrightarrow{\beta\text{-glucosidase}} \text{glucose}$$

Most of what is known about the enzymatic degradation of cellulose largely ignores the organism-substrate-relationship that is an integral part of the process. In reality our studies are confined to isolated enzymes and modified substrates. But there is no true alternative to using modified substrates. Thus, any mechanism of cellulose degradation should be approached from the actual fact of reactions. Let us, therefore, examine the case of this unique system in the context of the above mechanism and on the basis of actual data. Data (Ghose, 1969d) of $v-(S)$ for three cases of cellulose-cellulase are presented in Fig. 6 based on the data of Table 4.

When the reciprocals of v and (S) are plotted (Fig. 7) the expected linearity is shown in the three cases. The values of $1/V$ and $1/K$ are graphically evaluated.

The case of competitive inhibition (Aiba *et al.*, 1965) may be presented in the following scheme:

$$E + S \underset{K_2}{\overset{K_1}{\rightleftarrows}} ES \overset{K_3}{\longrightarrow} E + P, \tag{12}$$

$$E + I \underset{K_5}{\overset{K_4}{\rightleftarrows}} EI \tag{13}$$

and the empirical equilibrium constants for the Eqs. (12) and (13) may be identified as:

$$K_s = (S) \frac{e_t - e_{es} - e_{ep}}{e_{es}} = \frac{K_2}{K_1} \tag{14}$$

and

$$K_p = (p) \frac{e_t - e_{es} - e_{ep}}{e_{ep}} = \frac{K_5}{K_4}. \tag{15}$$

Where: K_1 = rate constant of the forward reaction between enzyme and substrate,

K_2 = rate constant of the reverse reaction between enzyme and substrate,

K_3 = rate constant of the forward reaction in the dissociation of the enzyme substrate complex,

K_4 = rate constant of the forward reaction between inhibitor and enzyme,

K_5 = rate constant of the reverse reaction between inhibitor and enzyme,

P = inhibitor,

K_s = equilibrium constant between E and ES,

K_i = equilibrium constant between E and I

and e_t = conc. of total enzyme,

e_{ep} = conc. of enzyme product complex,

e_{es} = conc. of enzyme-substrate complex,

p = conc. of product; and assumed that

$S \gg e_t,$

$p \gg e_{es},$

$V = K_3 e_t,$

$K_3 \ll k_1.$

5*

In this case,

$$v = K_3 e_{es} = \frac{K_3 e_t(S) K_p}{K_s K_p + K_s(p) + K_p(S)} \tag{16}$$

$$= \frac{V(S)}{K + (S) \cdot \dfrac{K}{K_p}(p)} . \tag{17}$$

Rearranging (17), to a linear form,

$$\frac{1}{v} = \frac{K}{V}\left(1 + \frac{p}{K_p}\right) \cdot \frac{1}{(S)} + \frac{1}{V} . \tag{18}$$

If inhibition is caused by product, then I may be replaced by P and K_i by K_p. The inhibition mechanism may be described as formation of a complex between the enzyme and the inhibitor. This results in the partial loss of the compatibility to form the product. In such a situation, therefore, the amount of product expected in the uninhibited reaction is always greater than that in the inhibited regardless of the nature of inhibition. The Eq. (18) should thus represent the model for competitive substrate or product inhibition. The rate of reaction should thus vary with either substrate concentration (in absence of inhibitor or product) or inhibitor (or product) concentration. It is not possible to evaluate the interdependence of the inhibition process between the three independent components: the substrates, the products and known inhibitors. The Cellulose-cellulase system is one of competitive inhibition by two products of the process. This follows from the similarity of values of equilibrium constants, K_p for both cellobiose and glucose. The reciprocal plots for no inhibition, glucose and cellobiose inhibition based on the data of Table 4 are presented in Fig. 7.

(a) For no-inhibition

$$K = 3.45 \text{ mg/ml}, \quad \text{and} \quad V = 1.3,$$

(b) for cellobiose inhibition ($p = 5 \text{ mg/ml}$)

$$K_{p\,cellobiose} = 28.09 \text{ mg/ml},$$

(c) for cellobiose inhibition ($p = 10 \text{ mg/ml}$)

$$K_{p\,glucose} = 24.21 \text{ mg/ml}$$

from Eq. (10):

$$K = \frac{K_2 + K_3}{K_1}$$

when $K_2 \gg K_3$, the substrate dissociation constant, K_s becomes:

$$K_s = K = \frac{K_2}{K_1}$$

and when $K_2 \ll K_3$, the Kinetic constant, K becomes:

$$K' = K = \frac{K_3}{K_1}.$$

Thus, for complete interpretation of the reactions it is necessary to have the values of K_2 and K_3 determined.

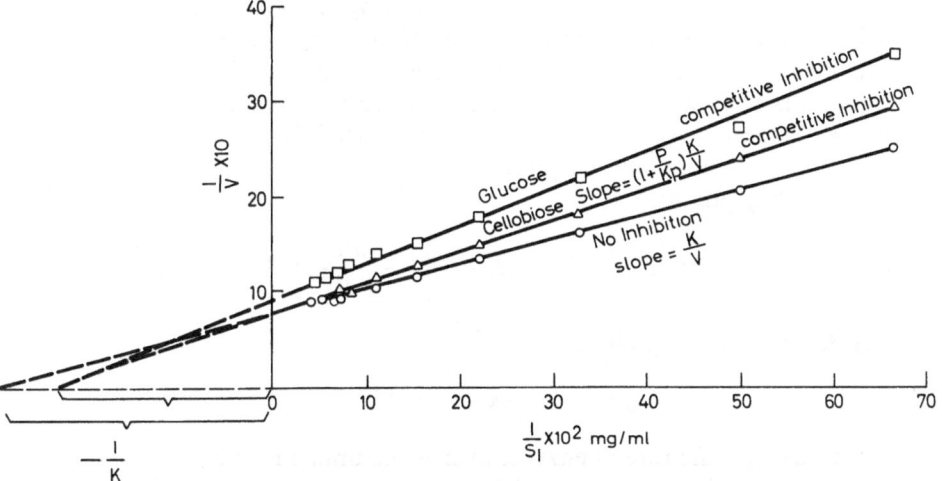

Fig. 7. Lineweaver-Burk plots for no inhibition, glucose and cellobiose inhibition in $< 25\,\mu$ cellulose, enzyme activity -2.0 FP, temp. $50°$ C

The use of K to represent the Michaelis constant (as distinct from the dissociation and kinetic constants) as determined from the kinetic data by graphical analysis, where the biological meaning is unknown and of K_s representing the true dissociation constant of the ES complex or substrate constant has been described in an earlier work (Dixon and Webb, 1958).

It is generally assumed that the substrate is activated after being coupled.

If it is maintained that a temporary abnormal electron configuration (implied in transition state kinetics) is the state of the activated substrate then the usual ES complex does not represent a transitional configuration

(Webb, 1963), but, rather, is a more stable species. However, the *ES* complex does pass through a truly activated state in the course of coupling, and, in the case of a solid system, through a chain of such activated states. In the case of cellulose-cellulase system products (cellobiose or glucose) seem to be unbound to the enzyme; but in others the products form complexes which are comparable to the *ES* complexes. Because cellulose is an insoluble substrate it has not been proven that there exists a product enzyme complex that is comparable to the enzyme substrate complex. The mechanism of the cellulose breakdown is an extremely complex system.

Process kinetic studies of α-amylase production by *B. subtilis* (Terui, 1968) and glucamylase formation by *A. niger* (Okazaki et al., 1965; Okazaki and Terui, 1966) have been shown to be compatible with the rate processes of acid protease production by *A. niger* (Shinmyo et al., 1969). Glucamylase forming system of *A. niger* showed a greater stability than α-amylase forming system of *B. subtilis*. The kinetic equations were applied to the acid protease producing system on the basis of several assumptions which were shown to be substantially valid. The two kinetic models are given below:

Growing phase:

$$\varepsilon = e^{-k(t-t_0)} \left\{ a \int_{t_0}^{t} \mu_g e^{k(t-t_0)} \, dt + b \int_{t_0}^{t} \left(-\frac{d\mu_g}{dt} e^{k(t-t_0)} \, dt \right) + \varepsilon_0 \right\}$$

and for non-growing phase

$$\varepsilon = \varepsilon_0 e^{-k(t-t_0)} + k_1 (e^{-\lambda(t-t_0)} - e^{k(t-t_0)}),$$

where ε = specific rate of enzyme formation, units $mg^{-1} h^{-1}$,

μ_g = specific growth rate, h^{-1},

k = rate constant for the decay of mRNA, h^{-1},

λ = rate constant for the decay of cellular RNA in non-growing phase, h^{-1},

ε_0 = ε-value at the starting time, t_0,

t = time, h,

t_0 = time at the start of experiment, h,

k_1 = is a constant.

These workers have also demonstrated the validity of the kinetic equations based on experimental data. However, the mechanism by which the enzyme-forming systems in derepressed cultures are not affected by repressing metabolites could not be explained. But, there exists a scope for verification of these models with the cellulose-cellulase system forming C_1 and C_x complexes.

5. Application of Chemical Kinetics and Evaluation of Rate Constants

Chemically speaking, product accumulation takes place as a result of higher rate of product formation than of product destruction. Let us examine the process of cellulose breakdown from a similar approach. The overall breakdown of cellulose can be grouped into two distinct steps, first depolymerization and hydrolysis causing the products to appear followed by a simultaneous disappearance of glucose due to side reactions of destructive nature. Accordingly,

$$(C_6H_{10}O_5)_n \xrightarrow[+nH_2O]{K_c} n(C_6H_{12}O_6), \tag{1}$$

$$p(C_6H_{12}O_6) \xrightarrow{K_s} p(CO_2 + H_2O). \tag{2}$$

Rate of cellulose saccharification is thus:

$$-\frac{dC}{dt} = K_c C_c. \tag{3}$$

Rate of glucose formation:

$$\frac{dC_{s_1}}{dt} = n\left(-\frac{dC_c}{dt}\right) = nK_c C_c \tag{4}$$

where $n =$ no. of moles of products formed.

Rate of glucose destruction is

$$-\frac{dC_{s_2}}{dt} = pK_s C_s. \tag{5}$$

Combining (4) and (5), overall rate of accumulation of glucose becomes:

$$\frac{dC_s}{dt} = \left(\frac{dC_{s_1}}{dt} - \frac{dC_{s_2}}{dt}\right) = nK_c C_c - pK_s C_s \tag{6}$$

where, concentration of accumulated glucose, gm-mole/l $= C_s$,
   initial concentration of cellulose, gm-mole/l $= C_{co}$,
   concentration of glucose decomposition products,
      gm-mole/l $= C_d$,
   amount of decomposition products, moles $= p$.

Thus, cellulose concentration, gm-mole/l,

$$C_c = \left(C_{co} - \frac{C_s}{n} - \frac{C_d}{p}\right). \tag{7}$$

The Eq. (6) can now be rewritten as:

$$\frac{dC_s}{dt} = nK_c\left(C_{co} - \frac{C_s}{n} - \frac{C_d}{p}\right) - K_s C_s .$$

Neglecting the amount of glucose decomposition product being very small, we get,

$$\frac{dC_s}{dt} = nK_c\left(C_{co} - \frac{C_s}{n}\right) - K_s C_s$$

$$= K_c\left[nC_{co} - C_s\left(1 + \frac{K_s}{K_c}\right)\right]. \tag{8}$$

Integrating we get:

$$\ln\left[\frac{1}{1 - \frac{C_s}{nC_{co}}\left(1 + \frac{K_s}{K_c}\right)}\right] = (K_c + K_s)\,t . \tag{9}$$

From the actual experimental conditions C_s may be computed as (assuming non-cellulosic fraction being small):

$$C_s = \frac{W_s}{180 \times 1000} \times \frac{1000}{20} ,$$

$$C_{co} = 0.4/n \times 162 \times \frac{1000}{20} .$$

Therefore, $\dfrac{C_s}{nc_0} = \dfrac{W_s}{444}$. Thus, the final form of the equation becomes:

$$2.303 \log\left[1 - \frac{W_s}{444}\left(1 + \frac{K_s}{K_c}\right)\right]^{-1} = (K_c + K_s)\,t . \tag{10}$$

Determination of K_s:

The rate of glucose destruction (Eq. (5)) on integration gives:

$$-\int \frac{dC_s}{C_s} = \int K_s\,dt$$

or

$$-\ln C_s = K_s t + I .$$

At

$$t = 0, \quad C_s = C_{so}; \quad \text{and} \quad I = -\ln C_{so}$$

or,

$$K_s t = \ln \frac{C_{so}}{C_s} .$$

A similar approach was made to evaluate the rate constant for the decomposition of glucose as functions of acid concentration and temperature in earlier work (Saeman, 1945).

Therefore,

$$K_s = \frac{2.303 \log \frac{C_{so}}{C_s}}{t}. \tag{11}$$

The data for glucose destruction rate in the case of acid hydrolysis of rice hulls using 45% H_2SO_4 and a glucose to acid ratio of 1 : 20 at 85° C was applied to the model to evaluate K_s. Two values of t for the same initial glucose concentration were selected and K_s evaluated (Das, 1969). Having obtained the av. value of K_s, the Eq. (10) can be solved for K_c on the basis of additional experimental data (Table 6).

Table 5

Contact time, t min	Initial glucose W_s, mg	Final glucose W_s, mg	K_s from Eq. (11), sec^{-1}
5	150	149.6	1.105×10^{-5}
20	150	148.1	1.087×10^{-5}

Av. value of $K_s = 1.096 \times 10^{-5}$ sec^{-1}.

Table 6

Contact time, t min	Glucose formed mg/gm of hull	Conversion into glucose		K_c sec^{-1} $\times 10^{-3}$
		As % of hull cellulose	As % of whole hull	
5	80.9	20.20	8.09	0.66
10	140.0	35.00	14.00	0.62
15	197.0	49.25	19.70	0.64

Av. value of $K_c = 0.64 \times 10^{-3}$ sec^{-1}.

6. Conclusion and Future Efforts

From the data on pure and crude cellulose saccharification presented in the paper it is evidenced that the initial enzymatic saccharification of cellulose is of pseudo first order and is therefore represented by

$$t = \frac{1}{K} \ln \frac{H_{co}}{H_{co} - H_c}$$

where t = reaction time, min,
   K = velocity constant, min^{-1},
   H_c = cellulose conc. g/ml,
   H_{co} = cellulose conc. initially, g/ml.

The presence of strong product inhibition and interference of the non-cellulosic components in the overall reaction results in the kinetics more complicated than in a soluble one phase enzyme-substrate system. The magnitude of activation energy (14,870 cal/gm mole; Das, 1969) estimated for rice hull-cellulase system falls in the range such that the reactions could be either a diffusion in aqueous phase or a chemical reaction. The involvement of large insoluble molecules suggests that both diffusional and chemical reaction processes are active.

The enzymatic saccharification of cellulose is a product inhibited system. The Linewearer Burk model is:

$$\frac{1}{v} = \frac{K}{V}\left(1 + \frac{p}{K_p}\right)\frac{1}{(S)} + \frac{1}{V}.$$

The model describes the system for the case of competitive product inhibition. This is verified from the calculated values of equilibrium constants K_p for the both cellobiose and glucose.

For acid hydrolysis the rate constants were found to be practically constant at the initial stages of product formation when product removal (destruction) is small. A simplified model for acid hydrolysis requiring product destructions can be given by

$$t(K_c + K_s) = 2.303 \log\left[1 - \frac{W_s}{444}\left(1 + \frac{K_s}{K_c}\right)\right]^{-1}.$$

It is essential to know more about the biological meaning of cellulose degradation in consecutive or simultaneous steps before a better evaluation of the kinetics can be possible. The data presented and the models devised therefrom may constitute a valid basis for further work. We do not believe that any approach formulating kinetics of cellulose hydrolysis based on data drawn from soluble and lower molecular weight systems as substrates can represent the correct situation. Information on substrate inhibition and the values of rate constants for C_1 and C_x activities are needed to advance beyond what has been presented here. Some information on adsorption isotherms for native and modified cellulose (C_1, C_x, mixtures of C_1 and C_x) and surface active agents which apparently increase activities will be of significant help in presenting a clearer picture on the kinetics of this important and complex enzyme system.

Symbols

A	Substrate surface destroyed
a	Initial substrate concentration, g/l
α	Rate constant for surface destruction

C_s Concentration of accumulated glucose, gm-mole/l
C_{co} Initial concentration of cellulose, gm-mole/l
C_d Conc. of glucose decomposition products, gm-mole/l
E Enzyme concentration, $mg\,ml^{-1}$
E_0 Initial enzyme conc., $mg\,ml^{-1}$
E_t, e_t Total enzyme conc., $mg\,ml^{-1}$
e_{es} Conc. of enzyme-substrate complex
e_{ep} Conc. of enzyme-product complex
ε Specific rate of enzyme formation, $mg^{-1}\,time^{-1}$
ε_0 Specific rate of enzyme formation at the starting time, to, $mg^{-1}\,time^{-1}$
H_c Conc. of hull cellulose saccharified
H_{co} Initial conc. of cellulose
K Michaelis constant, $mg\,ml^{-1}$
k Rate constant, $time^{-1}$
k_1 Rate const. of the forward reaction between E and S
k_2 Rate const. of the reverse reaction between E and S
k_3 Rate const. of the forward reaction in the diss. of ES complex
k_4 Rate const. of the forward reaction between P and E
k_5 Rate const. of the reverse reaction between P and E
K_c Rate const. for the forward reaction between cellulose and acid, $time^{-1}$
K_s Equilibrium const. between E and ES, or rate const. for the decomposition of glucose, $time^{-1}$
K_i Equilibrium const. between E and I
n Number of moles of product
P Inhibitor
p Product concentration
S Substrate conc.
S_0 Substrate surface available
t Contact time
t_0 Time at the start of experiment
v Steady state velocity
V_m Maximum velocity
μ Rate of hydrolysis
μ_g Specific growth rate, $time^{-1}$
λ Rate const. for the decay of cellular RNA in non-growing phase, $time^{-1}$
W_s Weight of glucose, mg

References

Aiba, S., Humphrey, A. E., Millis, N. F.: In: Biochemical Engg. Tokyo: Univ. of Tokyo Press 1965.
Alberty, R. A.: In: Adv. in Enzymology-Nord, F. F. (Ed), pp. 1—64. New York: Interscience Pub. Inc. 1956.
Amemura, A., Terui, G.: Proc. V. Symp. Cellulase and Related Enzymes, pp. 39—43; Dept. of Ferm. Tech., Osaka Univ., Japan 1965.
Briggs, G. E., Haldane, J. B. S.: Biochem. J. **19**, 383 (1925).
Chen, J. W., Koepsell, H. J., Maxon, W. D.: Biotech. Bioeng **4**, 65 (1962).
Das, K.: Ph. D. Diss. Calcutta: Jad. Univ. 1969.
Dawes, E. A.: In: Quantitative Problems in Biochem., p. 99—154. London: E. and S. Livingston 1962.
Dixon, M., Webb. E. C.: In: Enzymes, p. 69. New York: Acad. Press 1958.

Flora, R. M.: Ph. D. Diss., Va. Poly. Inst., Va. 1964.

Ghose, T. K.: Biotech. Bioeng. 11, 239 (1969a).

— Unpublished data (1969d).

— Kostick, J. A.: In: Adv. Chem. Series. Hajny, G. J., Reese, E. T. (Eds.), Am. Chem. Soc. 95, 415 (1969b).

— — Paper presented at the ACS Symposium on Industrial Microbial Enzymes. N.Y. City, Sept. 1969c.

Greenwood, C. T., Milne, E. A.: In: Adv. in Carbohydrate Chemistry. Wolfrom, M. L., Tipson, R. S. (Eds.), pp. 281—366. New York: Acad. Press 1968.

Halliwell, G.: Biochem. J. 95, 270 (1965).

King, K. W.: Biochemical and Biophysical Res. Comm. 24 (3), 295 (1966).

Mandels, Mary, Reese, E. T.: In: Adv. in Enzymic Hydrolysis of Cellulose and Related Materials. Reese, E. T. (Ed.), pp. 115. New York: Pergamon Press 1963.

— — Ann. Rev. Phytopathology 3, 259 (1965).

Mikhailov, N. V.: J. Poly. Sc. 30, 259 (1958).

Nisizawa, K., Hashimoto, Y., Shibata, Y.: In: Adv. in Enzymic Hydrolysis of Cellulose and Related Materials. Reese, E. T. (Ed.), pp. 290. New York: Pergamon Press 1963.

Okazaki, M., Shinmyo, A., Terui, G.: J. Fermentation Technol. (Japan) 43, 581 (1965).

— Terui, G.: J. Ferment. Technol. (Japan) 44, 276 (1966).

Reese, E. T.: Appl. Microbiol. 4 (1), 39 (1956).

— Levinson, H.: Physical. Plant. 5, 345 (1952).

Saeman, J. F.: Ind. and Engg. Chem. 37 (1), 43—52 (1945).

Shinmyo, A., Okazaki, M., Terui, G.: In: Fermentation Advances. Periman, D. (Ed.), pp. 337. New York: Academic Press 1969.

Stone, J. R., Scallan, A. M., Donefer, E., Ahlgren, E.: In: Adv. Chem. Series. Hajny, G. J., Reese, E. T. (Eds.), pp. 95, Am. Chem. Soc. 1969.

Terui, G.: J. Ferment. Technol. (Japan) 46, 427 (1968).

Walseth, C. S.: Tappi. 35 (5), 233 (1952).

Webb, J. L.: In: Enzyme and Metabolic Inhibitors. Webb, J. L. (Ed.), pp. 28. New York: Academic Press 1963.

Whitaker, D. R.: In: Adv. in Enzymic Hydrolysis of Cellulose and Related Materials. Reese, E. T. (Ed.), pp. 51. New York: Pergamon Press 1963.

— Colvin, J. R., Cook, W. H.: Arch. Biochem. and Biophys. 49, 257 (1954).

Wong, J. T., Hanes, C. S.: Canad. J. Biochem. Physiol. 40, 763 (1962).

T. K. Ghose and K. Das
Biochemical Engineering Laboratory
Department of Chemical Engineering
Indian Institute of Technology
New Delhi, India

CHAPTER 4

Production and Applications of Enzymes

W. T. Faith*, C. E. Neubeck, and E. T. Reese

With 2 Figures

Contents

* Died July 31, 1970.

The incorporation of enzymes into detergents is the most recent of a series of applications, and has aroused a great deal of interest in other possible uses. A relatively large number of earlier applications of enzymes did not have the benefit of modern day advertising and as a result did not capture the interest of the general population. Yet many of these applications retain their place in industry because of their unique nature.

The concept of using enzymes in industrial applications dates back to 1894 when a patent was obtained (Takamine, 1894) for making diastatic enzymes from fungi. Unfortunately, formation of off flavor hampered use of these early preparations in alcohol production (Takamine, 1914). A few years earlier it was claimed (Röhm, 1908) that pancreatic enzymes could be used in the bating of hides, and these preparations were quickly accepted in the leather industry. Three years later, a process was patented (Wallerstein, 1911) for the chill-proofing of beer using proteolytic enzymes, and again this process was quickly accepted. The use of tryptic enzymes as laundry aids (Röhm, 1913) was not widely accepted although it was used in Europe. The contributions of these three men mark the beginning of industrial enzymes, and of companies that have remained leaders in this field.

Several other applications of enzymes into the commercial area followed. The use of pectinases for the clarification of fruit juices was introduced in the early 30's both in Germany (Mehlitz, 1930) and in the United States (Kertesz, 1930). The use of fungal enzymes for the preparation of sweet sirups (Dale and Langlois, 1940; Langlois, 1940) developed under the stimulus of the sugar shortages of World War II. This development has resulted in a large variety of sirups, and modifications are still being made. The introduction of enzymes, other than from cereal grains, to the baking industry in the early 50's followed much experimental work at Kansas State University (Johnson and Miller, 1949) using enzymes supplied by the commercial producers. Interest in the use of fungal enzymes in the distilling industry, as suggested earlier by Takamine, was revived in the World War II era because of the pressing need for alcohol, but it required the commercial development of fungal amyloglucosidase in the last decade to make the use of microbial enzymes more attractive than malt in this industry.

The present chapter is limited to presenting developments in the enzyme field related to commercial practices. We have attempted to indicate the variety of commercially available enzymes and procedures used to produce them. The extent to which microbial enzymes have enhanced or replaced the enzymes derived from plants and animals is indicated.

A. Research

1. Selection of Organism

The selection of an organism is the first important step in the development of any product. For microbial enzyme production several factors must be considered:

a) Location of Enzyme

Cell-bound enzyme has an advantage in that it is concentrated in tissue which can be readily separated from the fermentation broth. But being cell-bound, it involves the difficulties associated with disintegration of cells (Edebo, 1969), and subsequent separation from other cellular components (Lilly and Dunnill, 1969). Though many enzymes are never released by living cells, (i.e. the "cell-bound" category), relatively few of the commercial enzymes (e.g. yeast invertase) are produced in this manner.

Secreted enzymes, on the other hand, must be concentrated from very dilute solutions of culture filtrate, a process which may be costly. In general, their "purity" is appreciably greater than that of enzymes present in cell-extracts. Most commercial enzymes are of this type.

b) Nature of Organism

The preferred organisms are those which "handle" easily or, in the case of food applications, have the FDA stamp of approval. They are stable in their characteristics of enzyme yields, spore production, and cultural requirements. They present minimal difficulty in filtration, centrifugation, or disintegration (if required). They produce no toxic or other undesirable side products, and they grow on inexpensive substrates.

c) Yield and Purity of Enzyme

Purity is generally measured as specific activity (enzyme per unit weight of protein), which may be quite misleading when polysaccharides, polyesters, etc. are present in large amounts. As a consequence industrial enzymes are usually measured in terms of activity per unit weight. Currently it appears that a reasonable goal is the obtention of a culture filtrate, or cell extract, in which the desired enzyme makes up 10% of the total protein. One organism has been reported to produce a particular enzyme in an amount equal to one third of its total protein

(Table 1). When that enzyme is β-galactosidase, one wonders whether the organism or the investigator is in error. There are many examples of enzymes which require 1000 (+) fold purification, but relatively few in which "purity" is obtained by a mere 10-fold increase in specific activity (Table 1). Rather unexpectedly, some of these high purity enzymes are found in cell extracts and so represent a large percentage of all protein synthesized (Pardee, 1969). Secreted enzymes represent much less of the total cell protein.

Table 1. "*Purity*" *of enzymes in culture filtrates or cell extracts*

Enzyme	Source	R^a	Ref.
Aspartate transcarbamylase	E. coli	12	Gerhart and Holoubek (1967)
Catalase	Rhodopseudomonas spheroides	4	Clayton and Smith (1960)
β-galactosidase	E. coli	16	Wang and Humphrey (1969)
β-galactosidase	E. coli	3	Novick and Horinchi (1961)
Exo-β1,3 glucanase	Basidiomycete sp. QM 806	7	Huotari et al. (1968)
Exo-β1,3 glucanase	Sclerotinia libertiana	8	Ebata and Santomura (1963)
Tyrosinase	Neurospora crassa	20	Horowitz et al. (1961)

$$^a \; R = \frac{\text{Spec. Act. "Pure" Enzyme}}{\text{Spec. Act. "Crude" Enzyme}}.$$

Selection of an organism cannot, of course, be independent of the conditions involved in the screening procedure. An organism selected for amylase production by screening in surface culture or bran medium may be quite different from that screened in shake culture on starch. Although the trend in enzyme production is away from surface culture, certain enzymes are best produced by it, and there is difficulty producing these enzymes in any other way. Enzymes designed to be used under severe conditions, at high pH, high temperature, etc., must be screened under similar adverse conditions.

A very necessary factor in any screening program is the availability of large numbers of diverse organisms. In addition to the government supported collections (USDA, Peoria, Ill.; U.S. Army Lab., Natick, Mass., and American Type Culture Collection, Rockville, Md.) large companies maintain their own cultures. Frequently investigators find it advantageous to isolate organisms by enrichment procedure from natural sources such as soil or water. Based on the organisms available and the screening procedures used, it is not at all unusual that different investigators develop enzyme processes based on different organisms (e.g. for amyloglucosidase: *Aspergillus niger, A. awamori, A. phoenicis, Rhizopus delemar, R. oryzae, Endomycopsis sp.*).

2. Mutation

The selected organism may further be improved through mutation. Table 2 illustrates the type of increases reported in published work. Techniques have been described for obtaining (a) mutants resistant to catabolite repression, (b) constitutive mutants which produce enzyme in the presence of repressor, and (c) constitutive mutants which form enzymes without the addition of inducer (Demain, 1968). The aim of most work along these lines has been the increased production of relatively simple compounds, e.g. aminoacids and antibiotics. The procedures involve resistance to antimetabolites (analogs of the desired product) as a means of selecting mutants. For the production of hydrolytic enzymes, there seems to be no similar simple means for a preliminary selection of the desired mutant, unless enzyme production happens

Table 2. *Increases in enzyme production by mutation*

Enzyme	Organism	Enzyme Yield $\frac{(\text{Mutant})}{(\text{Control})}$	Ref.
Aspartate transcarbamylase	E. coli	500	Gerhart and Holoubek (1967)
Cellulase	Trichoderma viride QM 6a	2	Mandels et al. (1970)
Dihydrofolic reductase	Diplococcus pneumoniae	200	Sirotnak et al. (1964)
Dihydrofolic reductase	Streptococcus faecalis	10—100	Hillcoat and Blakley (1966)
β-galactosidase	E. coli	4	Novick and Horinchi (1961)
Glucoamylase	Aspergillus foetidus	1.6	Underkofler (1970)
Homo-serine dehydrogenase	E. coli	3	Egorov et al. (1965)
Protease	Bacillus cereus	10	Levisohn and Aronson (1967)
α-amylase	Aspergillus oryzae	10	Neubeck (1970, unreported)

to be correlated with the yield of some simple product. In some instances, a mutant lacking the desired enzyme can be reverted, yielding a culture giving much more enzyme than its grandparent (Egorov, 1965). This again is a procedure which has been more amply demonstrated on products other than enzymes, but which may prove increasingly valuable. The frequency of reversion (using UV) increases with percent kill. At 99[+] % kill as many as 6% of the survivors are revertants (Dulaney and Dulaney, 1967).

Significant increases in the enzyme productivity of organisms used commercially have been achieved using ultraviolet light and nitrogen mustards as mutagens. Frequently alternate treatments with a variety of mutagens have been more beneficial than a series involving the same mutagen. The original strains of *A. oryzae* used in the production of commercial amylase in the early 40's show only one-tenth to one-fifteenth the productivity exhibited by the presently available mutant organisms[1]. The new organisms show significant morphological changes (colour; extent of sporulation) from the original strains. Attempts to demonstrate the increased enzyme activity on starch agar in simple zone tests have not been successful. Selection of the improved organisms was made almost entirely on the basis of screening tests carried out on complex media, using the survivors remaining after mutagenic treatment. Similar results have been obtained with other enzymes and the organisms which produced them (Mandels *et al.*, 1970).

3. Modification of Conditions of Growth

a) General Methods

Enzymes are produced in three ways which may be classified as bran culture, submerged culture, and two step submerged culture.

The surface bran culture method involves growth of the organism on moist acidified bran (either from wheat or rice) in beds through which air is circulated. Bran itself is a good nutrient for amylase production, but the bran may be fortified with other nutrients and salts to improve yields. In another modification air may be passed over the bed. The surface procedure has also been called the koji method when rice is used as the substrate. The terms tray or semisolid culture method are also used. Usually the beds are thin layers one to two inches deep to expedite heat removal. An excellent review of the tray process was published in 1947 (Underkofler *et al.*, 1947). Except for slight modifications, the process is still carried out commercially. More recently, the Japanese (Terui and Takano, 1960) have investigated on a commercial scale a

---

[1] Unpublished experiments by C. Neubeck.

deep bed bran (and also koji) process using layers of medium several feet thick and several feet in diameter. This process is called the "high heap" method. Another modification of the moist bran process involves tumbling of the growing culture in rotating drums to obtain the required aeration and process control. This method was tested extensively in the laboratory in the 1930–1940 period (Underkofler *et al.*, 1939), and one commercial producer of enzymes is using it. The bran procedures are most widely used for growth of fungi, and the term "moldy bran" is frequently applied to the crude dry product arising from any of the procedures.

The deep tank or submerged process involves growth of the microorganism in liquid culture within large vessels equipped for aeration and agitation. The tanks are the same as those described for typical submerged fermentation processes. It is the more widely used method now for producing enzymes, because it is easier to control than the moist bran methods and the growing medium can be varied more easily. The process is usually a batch system, but it may be adapted to continuous

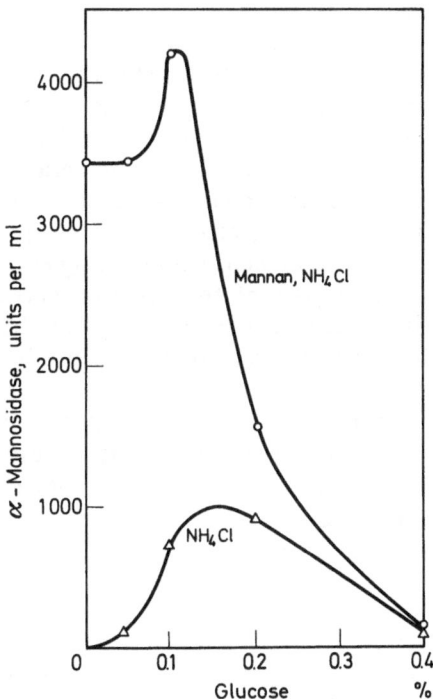

Fig. 1. Effect of glucose concentration on yields of α-mannosidase by washed mycelium (Demain)

6*

fermentation or modified to the extent that supplemental nutrients are added to maintain growth and activity production.

The third procedure is a two step operation in which microbial cells produced in a fermenter are transferred to a reaction vessel in which the conditions differ from those used in growth. α-Mannosidase has been produced in this way (Inamine *et al.*, 1969). *Streptomyces griseus* cells were grown in a rich medium for 17 hours at 28° C, washed, and transferred to a dilute mineral salts solution containing an inducing compound (usually yeast α-mannan). Maximum α-mannosidase concentration, obtained in 18–24 hours, was accompanied by little, if any, growth. This enzyme is useful in converting mannosido-streptomycin to the more active antibiotic form of streptomycin. Advantages of the method include reduced requirement for inducer, high yield of enzyme, and relatively small amounts of contaminating nutrients. The method takes advantage of the fact that conditions for growth (trophophase) differ from those for product elaboration (idiophase). Separation of the phases permits closer control of the optimal conditions for each. The authors are not aware of commercial implementation of this technique for the production of enzymes.

As in the production of many metabolic substances, conditions optimal for growth are not necessarily those which give maximum enzyme yields. Growth is essential, but in general, high concentrations of readily utilizable nutrients should be avoided. Close attention must be given to control of pH, mineral content, temperature and aeration. The growing conditions and nutritional requirements must be worked out empirically for each organism for the selected method of growth. The interrelationships between the several factors influencing enzyme production are very complex. A large part of the research carried out by the industrial producer is concerned with working out the most efficient method of growing the microorganism for enzyme production and translating these results to the plant production units.

Selection of the method of growth will depend on how well the surface or submerged methods can furnish the aeration, temperature and pH control, and supply of nutrients and inducers required for production of the desired enzyme.

b) Inducers

Many organisms are stimulated to increased enzyme synthesis by adding an inducer to the culture. The inducer was formerly considered to be the substrate of the enzyme or a modification of the substrate. More recently it has been shown that products (and modified products) of the enzyme may also be inducers, e.g. of polysaccharases. This would appear to be a necessary consequence of the inability of polymers to enter the cell. As a result it seems likely that this principle will also apply to induction of other depolymerases, such as proteases and nucleases.

Failure to recognize induction by products of enzyme action resulted from the well-known catabolite repression. Repression occurs at relatively low concentration of the sugar dimers (which are the inducers of the corresponding polysaccharase). Thus at a level of 0.5% cellobiose very little cellulase is produced by organisms growing in shake flasks. However, continuous feeding of very low cellobiose concentrations (< 0.05 mg/ml) greatly increases cellulase yields (Suzuki et al., 1969). The inducing ability of a polymer (the enzyme substrate) results from its hydrolysis to dimer in these systems. The dimer never accumulates to the concentration at which repression occurs, because it is consumed by the organism as rapidly as it is formed.

Any method which can supply the inducer slowly to the growing organism will reduce catabolite repression and result in enhanced enzyme yields. Esters of the inducer have the necessary properties, provided that the organism possesses an appropriate esterase. The esters of inducing disaccharides (Table 3) are considerably more potent than the disaccharides themselves, presumably because the esters do not repress. Glycosides of the dimers may possibly induce, but ethers, because of their resistance to breakdown, would not be expected to do so.

Just how much increase in enzyme yields might one anticipate from the use of modified inducer? After it was found that sucrose mono-palmitate increased invertase yields as much as 100-fold (Reese et al., 1969), a screening of organisms was undertaken using this ester as substrate. Under these conditions, Pullularia pullulans was found to be an excellent source of invertase, and, unlike yeast, it secreted the enzyme into the culture filtrate. The yields are surprisingly good. The invertase contents of the unconcentrated filtrates have the same magnitude as that of a widely-distributed commercial concentrate. The specific activities are also quite similar (Reese, unpubl. data).

Similarly, adenosine is a true inducer of purine nucleosidase. But at a level of 0.5% its repressing effect is great enough to prevent the appearance of enzyme. Use of ribonucleic acid (RNA) or of the phosphate ester of adenosine (AMP) by the fungus leads to a low but continuous supply of adenosine and a resultant high enzyme concentration (Reese, 1968).

In apparent contradiction of the above, it is possible to obtain good enzyme yields at concentrations of an inducer that normally repress. This is achieved by decreasing the rate of consumption of the inducer, (a) by some means of inhibition, such as lower-than-optimal temperature;

(b) by use of compounds at concentrations which are slightly toxic; or
(c) by restricting one or more of the factors required for growth. It thus
appears that catabolite repression is not a matter of catabolite concentration, but of the rate at which the inducer is metabolized. An
alternate procedure, when there is a family of inducers, is to select the

Table 3. *Effect of modified inducers on enzyme yield*

Enzyme	Organism	Inducer	Yield	Ref.
Cellulase EC 3.2.1.4	*Trichoderma viride*	Cellulose	22.5[a]	Reese *et al.* (1969)
		Cellobiose	0.2[a]	
		Cellobiose dipalmitate	4.8[a]	
	Pestalotiopsis westerdijkii	Cellulose	35.9[a]	
		Cellobiose	0.2[a]	
		Cellobiose octaacetate	20.1[a]	
	Pseudomonas fluorescens	Cellulose	514	Suzaki *et al.* (1969)
		Cellobiose (slow feeding)	200	
		Sophorose	397	
Dextranase EC 3.2.1.11	*Penicillium funiculosum*	dextran	1080[a]	Reese *et al.* (1969)
		isomaltose	2[a]	
		isomaltose dipalmitate	1098[a]	
Invertase EC 3.2.1.26	*Pullularia pullulans*	sucrose	1.3[a]	Reese *et al.* (1969)
		sucrose monopalmitate	108[a]	
Purine nucleosidase EC 3.2.2.1	*Aspergillus ambiguus*	Adenosine	0	Reese (1968)
		Yeast RNA	57	
		Adenosine 5'PO$_4$	90	

[a] Unit values in International units. Others as defined by authors.

more slowly consumed member. Xylidine (Fahraeus, G., pers. comm.)
induced much more laccase in *Polyporus versicolor* than did other,
more rapidly consumed substrates, and sophorose ($\beta 1 \rightarrow 2$ diglucose)
is a better inducer of cellulase than cellobiose ($\beta 1 \rightarrow 4$ diglucose) for some
organisms (Table 3). In a similar way the D-isomers (the *un natural* forms)
of tyrosine and phenylalanine are considerably more active inducers of
tyrosinase in Neurospora than are the natural forms (L-isomers) of the
same aminoacids (Horowitz *et al.*, 1961). In some cases, the toxicity of
the agent may modify its inducing effect.

c) Surfactants as Promotors of Enzyme Production

Surfactants (esp. Tween 80)[2] at low concentrations have long been known to enhance the growth of some bacteria. Their effects in increasing enzyme yields have been reported recently (Reese and Maguire, 1969), although it is quite possible that the effect was recognized earlier and kept a trade secret. It may be that the sometimes observed increase in yields in tanks (vs. shake flasks) is due to the addition of antifoam agents (i.e. surfactants). More frequently, anti-foams depress the dissolved oxygen in a growing culture and produce poor activities.

Table 4. *Effect of addition of surfactant (Tween 80) to culture medium on enzyme yields (Reese, unreported)*

Enzyme	Source	$R^a \dfrac{\text{(Yield + Surfactant)}}{\text{(Yield − Surfactant)}}$
Cellulase	Many fungi	20 x
Invertase	Many fungi	16
β 1,3 glucanase	Many fungi	10
β-glucosidase	Many fungi	8
Xylanase	Many fungi	4
Amylase	Many fungi	4
Nucleosidase	Many fungi	5
Esterase	Many fungi	6
Dextranase	*Penicillium funiculosum* QM 474	2
Pullulanase	*Aerobacter aerogenes* QM B 1591	1.5

[a] R = Ratio of yield in shaken flasks (29°) containing appropriate culture media plus Tween 80 (0.1 %), to yield under identical conditions lacking Tween.

The addition of surfactants to culture media increases the yields of many enzymes. The results presented in Table 4 were obtained with secreted enzymes. No data are available for cell-bound enzymes. In some tests, no enzyme is detectable in the controls, but several units of activity per ml can be found in the presence of surfactants (i.e. infinite enhancement). The values shown in Table 4 are based on the surfactant Tween 80 at 0.1 %. For some systems, higher Tween concentrations give better yields, but there usually is a critical upper level. For some systems, other (nonionic) surfactants such as Triton[3] are superior to Tween. In general, the enhancement factor is greatest for organisms which do not normally secrete much enzyme and least for those organisms already selected for their high yields. But even in the latter, an appreciable

[2] Tween is a trademark of Atlas Chemical Co.
[3] Triton is a trademark of the Rohm and Haas Company, Phila.

increase may be anticipated. A doubling in cellulase was observed with the active cellulolytic fungus, *Trichoderma viride*, and in dextranase with the highly productive *Penicillium funiculosum*.

The mechanism by which surfactants increase enzyme secretion is not known. It seems likely that the "leakiness" of cell membranes may vary from one organism to another and thus account for the variability between strains in the amount of enzyme normally secreted. Surfactants may be expected to accumulate at the cell membrane and thereby further increase (or modify) leakiness. As the amount of cell-bound enzyme appears to remain constant (feed-back control), increased secretion leads to increased enzyme production as the cell attempts to maintain the cell-bound level. In the case of surfactant-stimulated production of α-amylase by *A. oryzae*, the higher enzyme production may be accompanied by a lower mycelial weight. At some critical level of surfactant, the effect on the cell wall becomes so great that the cells lyse, and growth as well as enzyme production cease.

An alternate explanation offered for the increased yields is that the surfactant protects the enzyme from inactivation. Thus, Tween protects laccase from inactivation by phenol (Fahraeus, G., pers. comm.), but this effect has not been demonstrated for many of the observed increases.

The enhancement of enzyme yields by carbohydrate esters (Table 3) may be due, in part, to their surfactant properties. This is certainly true of sucrose monopalmitate, which has been found to increase yields, not only of invertase, but of many other enzymes listed in Table 4. Both Tween 80 and sucrose monopalmitate are nonionic surfactants relatively nontoxic to the organisms tested. At first, it was believed that only this type of surfactant possessed the stimulating effect. Later sodium oleate proved superior to Tween 80 for producing cellulase by some fungi. The anionic surfactants tend to be more toxic than the nonionics. The cationic agents, which are even more toxic, have not been found useful in this work.

4. Immobilization

Most enzyme reactions take place in a solution of enzyme and substrate in a batchwise process, at the completion of which the enzyme is discarded. As the activity of the enzyme may be scarcely diminished in reactions carried out under optimum conditions, this practice is very wasteful. Immobilization of enzyme by various means prevents its diffusion and permits its later separation from the reaction mixture by simple filtration. Enzyme can be recovered in this way and reused many times, greatly improving the economy of operation. The easy removal of enzyme from the treated product may be a distinct advantage in food industry applications. In batch operations the processer usually depends on subsequent processing steps for removal or inactivates the enzyme in some other manner.

Enzyme localised on columns can be used in a continuous single reaction, or by successive layering of different enzymes, a series of reactions can be carried out, with the final product emerging in the effluent. In like manner, enzymes may be localized in animal tissues to carry out a required action.

Four means of localization have been described.

a) Microencapsulation
(Chang *et al.*, 1967; Chang and Poznansky, 1968)

Semipermeable microcapsules of cellular dimensions (10–20 μ dia 200 Å thick) can be formed, each containing enzyme, or some other large molecule. Useful membranes have been made of collodion or other polymeric materials. These membranes are permeable to the substrates and products of enzymic action, but the enzyme itself is too large to leak out. A great many enzymes have been successfully encapsulated, and the method appears to have general applicability. A limitation appears, however, in that large substrate molecules (proteins, poly-saccharides) cannot permeate into the capsule, and the system is not applicable to such hydrolyses.

b) Covalent Bonding of Enzyme to Insoluble Carrier
(Silman and Katchalski, 1966)

Enzymes have been bound to insoluble cellulose derivatives by various methods such as (a) to insoluble carboxymethyl cellulose using the azide derivative, (b) to insoluble cellulose using the diimide reaction, etc. More recently[4] H. Weetall described the covalent coupling of 1-amino-acid oxidase to porous silica glass particles.

The enzymes must, of course, be linked at some distance from the active site of the enzyme. Such insolubilized enzymes retain their activities, but their immobility may reduce the reaction rate. The properties of such covalently linked enzymes are not always the same as those of the free enzyme. The anionic or cationic nature of the carrier may alter the pH optimum for the reaction (Goldstein, 1969). The binding of the enzyme to carrier may result in steric hindrance and impose restrictions on the specificity of the bound enzyme. This is most

---

[4] ACS 1969 Meeting.

apparent where the substrate is a large molecule, such as protein, rather than a small molecule like a peptide. The products of protein hydrolysis may differ, those products from bound enzyme being of larger size than those from soluble enzyme. K_m and V_{max} values may also be affected.

c) Localization of Enzyme in the Aqueous Phase
of a Two-Phase System (Reese and Mandels, 1958)

An enzyme can be dissolved in an aqueous phase and retained on a column of a hydrophilic solid such as cellulose. Substrate in the solvent phase diffuses into the aqueous phase where reaction with the enzyme occurs. Products diffuse back into the mobile phase and pass out of the column. Using invertase, such a system retained nearly complete activity for a number of weeks. This system resembles the micro-encapsulated enzyme procedure, except that a solvent phase substitutes for a semipermeable membrane.

d) Retention of Enzyme by Ultrafiltration Membranes

In the preceding methods, the enzyme is confined to small droplets, capsules, or inert carrier. Enlargement of a capsule to the size of a fermenter is theoretically possible, and this principle has been applied with practical modifications. The enzyme is free in solution as in the batch procedures, but an ultrafilter serves to separate the reaction products from the enzyme and substrate. Substrate is continuously pumped into the system, and product is removed by ultrafiltration to provide a continuous system. This method is applicable only to systems where the substrate is a large or an insoluble substance, so that it – with the enzyme – are retained inside the membrane. Successful demonstration of experimental runs on cellulose saccharification (Ghose and Kostick, 1969) and on starch hydrolysis (Butterworth et al., 1969) have been made. Success depends upon the availability of suitable membranes and practical application on their cost.

All four means of immobilization have been used successfully in laboratory operation. Machinery has been developed for large-scale production of microcapsules with various contents. Chemical proce-dures have been described for covalent bonding of enzyme to carrier. It has been reported (Anon, 1970) that Tanabe Seiyakii Co., Osaka, is now using immobilized enzymes commercially to catalyze the hydro-lysis step in the production of D or L amino acids. It has also been reported (Takasaki et al., 1969) that glucose isomerase can be retained by *Streptomyces* cellular structures by heating and that this preparation can be used for a series of isomerizations. The preparation apparently diminishes in strength with the number of passes.

It seems likely that the use of penicillin-amidase supported on cellulose by chemical coupling with triazines will soon be employed commercially to convert penicillin to other forms. The use of amyloglucosidase supported on a cellulosic support also appears to be a good possibility for the near future.

B. Commercial Enzymes

1. Preparation of Enzymes for Industrial Use

The preparation of marketable enzyme products is essentially the same for enzymes prepared from different sources (plant, animal, microbial) since all are protein. The precise details vary and the use for which the product is prepared usually determines the number of processing steps and degree of purification.

The basic steps and their variations involve:

1. Solution of the above enzyme in water by extraction from the crude enzyme-containing material. Some very crude products simply represent a dry form of the enzyme-producing cellular structure and even this first step may be bypassed. In some cases an autolysis or activating step may be required prior to the water extraction.

2. Removal of cellular debris and other insoluble material from the aqueous enzyme extract using centrifugation, filtration, or both.

3. Precipitation of the enzyme from the aqueous solution by protein precipitants. The precipitate is then dried in air or *in vacuo*. The enzyme solution may be concentrated prior to the more selective precipitation step.

Fig. 2 is a diagram indicating some treatments which have been used in the preparation of enzymes on a commercial scale. Most of the equipment can be found in food-processing plants. Nearly all processing operations (except drying) are carried out at low temperature (0–10° C) to minimize denaturation of the enzyme. Enzyme products may be removed at several points to obtain a series of products differing in absolute activity and purity. The same enzyme source may be the starting point for a large variety of different products. The simplest products representing dried cellular material are used in such operations as the bating of hides. Liquid products are employed in textile starch desizing and fruit juice clarification. High activity products are used in fruit juice clarification and several other food-processing applications discussed below.

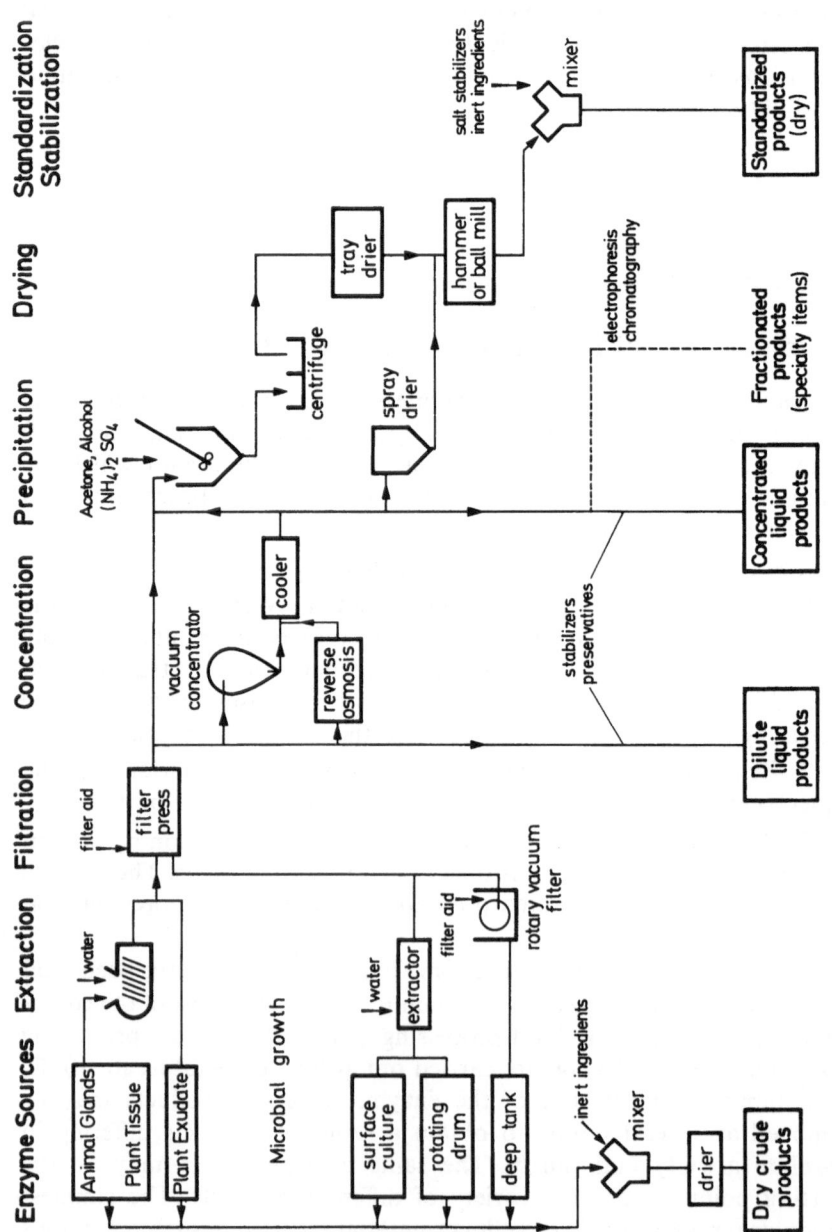

Fig. 2. Preparation of commercial enzymes

A listing of the media ingredients, filtration aids, precipitating agents, and formulating materials used by major enzyme producers in the preparation of microbial enzymes for food processing was presented by representatives of the industry (Beckhorn, 1965). A large variety of ingredients and process variables were required to encompass the various products prepared by the three manufacturers represented (Wallerstein Co., Rohm and Haas Company, Miles Chemical Co.).

The broad variety of different enzyme products has evolved because of the many applications in which they are being used. Enzymes are used in industrial applications which differ greatly in production volumes and degree of sophistication. Very active materials require very precise measurements and very effective mixing for efficient use. The small manufacturer has difficulty in using such products, and diluted standardized products fit his needs better. Automated industrial processes frequently utilize liquid products, because they lend themselves to pumping operations more readily than solids. It is often less expensive to prepare a liquid product and stabilize it against microbial spoilage and enzyme denaturation than to concentrate, precipitate, and dry the product for the user to redissolve. Unfortunately some enzymes are too labile to be handled readily in this way and they must be provided in dry form.

The various enzyme products are marketed on the basis of activity. Sometimes the activity is determined by a standard enzyme assay, but more frequently the activity is determined by methods based upon the industrial application of the enzyme. Large consumers with well-equipped laboratories find it expedient to purchase concentrates, but smaller plants prefer to purchase enzymes at standardized activities. Materials compatible with the intended application are added to the enzyme products by the enzyme manufacturer to adjust the activity to the required value. Stabilizers, activators and other materials to improve the response of the enzyme are also added. In the case of enzymes for food applications, the usual bacteriological tests (e.g. absence of Salmonella, gas formers, etc.) are applied to the products.

Most commercial enzymes in wide demand are not extensively fractionated. However, in some applications (baking, dextrose production), the presence of contaminating enzymes must be very low or rigidly controlled. One advantage of microbial enzymes results from the fact that an organism can be selected and grown under conditions where the level of contaminating enzymes can be kept low. In applications where high purity enzymes are required (e.g. pharmaceutical syntheses) the procedures required fall outside of the usual commercial practice. At the present time, chromatographic and electrophoretic techniques for the preparation of these high purity enzymes are restricted to pilot plant or laboratory scale where better process control can be obtained. The cost of such operations puts high purity products in the speciality rather than in the industrial class. The next five to ten years will probably see more extensive exploitation of these techniques in the commercial field as the demand for new and more highly purified enzymes arises. Some use of preferential adsorption has been made in the removal of trans-glucosidase from fungal amyloglucosidase. Many adsorbents and techniques have been disclosed in patents for this purpose. Among these are cellulose (Kerr, 1927), clays (Inglett, 1963), lignin (Cayle, 1923; Hurst and Turner, 1962), anionic ion-exchange resins (Corman, 1967) and Sephadex (Garbutt, 1967).

2. Comparison of Industrial and Laboratory Preparations of Enzymes

The simple flow diagram given for the preparation of industrial enzymes could serve as well to illustrate operation on a laboratory scale. The magnitude of the commercial operation, however, brings about critical differences between plant and laboratory procedures. Commercial producers of enzymes find it necessary to carry out laboratory and pilot plant work to develop the most efficient series of steps to be followed in the manufacture. Usually a process is developed from a microbial source, because animal and plant tissue represent a limited and essentially uncontrolled supply of raw material.

It was noted above that the initial step in preparing a microbial enzyme is the selection of the proper organism, the method of growth, and the cultural conditions required for maximum activity. These screening techniques must also conform to procedures which have utility in scaling up to equipment that the producer has available or can be engineered at practical cost. Ordinary flasks or jars, either shaken or stationary, are generally used for the growth at this stage using 50–1000 g of medium. The exact procedures, i.e. media, agitation conditions, temperature profile, etc., followed by the commercial producer are usually considered to be proprietary. The effects of the various screening techniques must be empirically determined for each new product, and usually some adjustments are necessary as mutants of greater productivity are developed. The next step in the transfer of the process to plant equipment must be carried out in large laboratory or pilot plant equipment (e.g. 5–100 gallons capacity) which can then be scaled up to the available plant equipment. The pilot equipment is operated at practical levels of such variables as power consumption, aeration, temperature control, medium sterilization etc. Agitation and aeration requirement during scale-up are discussed in other chapters of this volume, and these will not be discussed here except for the observation that submerged fungal cultures used for enzyme production are highly aerobic and tend to become very viscous. Scale-up, therefore, can be a difficult problem. The maintenance of a pure culture and proper growth temperature in commercial enzyme production deserves some additional comment here, because the control of these factors on a commercial scale may be very critical.

a) Maintenance of Pure Culture in Commercial Scale

Laboratory media in small quantities (e.g. 100 g of either submerged culture or surface culture) can be readily sterilized and maintained without contamination by simple closures of cotton or use of a sterile air supply, provided that the inoculum is handled properly. Under plant conditions, the sterilization of 1000 pounds or more of material requires adequate distribution of heat throughout the mass of medium and the preventation of leakage around rotating shafts, inoculation and sampling ports. Frequently the extent of contamination in raw materials is much higher in bulk, and this may be a problem because of longer make-up time. The sterilization of liquid media is more readily accomplished than the high solids bran media, but some degradation of medium is usually obtained in media of either the submerged or surface type. Maintenance of pure culture in the surface method is further complicated by the exposure of a large surface area (1000 square feet or more) to contamination. Control of such a large area over the usual two to four day growing period with sterile air is a formidable task. Protection from contamination in the initial stage depends on acidification of the bran. Rapidly growing organisms can effectively prevent contamination by other organisms by establishing a surface barrier, but this is not always the case. The rotating drum method using bran has many of the advantages of the submerged culture method as far as maintenance of pure culture, however, mycelium fragility and compaction during rotation can pose a problem in large masses of culture.

b) Maintenance of Proper Growth Temperature

The temperature of small quantities of growing culture can be readily controlled in the laboratory by immersing the growth vessel in constant temperature water baths or in rooms with good temperature control and high ratio of air change. Laboratory equipment can be operated over a wide range of temperatures. Changes in temperature can be made in a matter of minutes. Large equipment must rely on cooling jackets or coils for temperature control in submerged culture. Surface culture requires water-cooling or circulation of air to control the temperature. The amount of heat generated in the growing process for either method is very large. Essentially all carbohydrate in the medium is consumed with the evolution of CO_2, H_2O and heat. A typical 2500-pound surface culture charge containing about 1000 pounds of solids ($\simeq$ 400–600 pounds of starch) can produce enough heat during the most active stage of growth (from 20–40 hours) to raise the temperature of the entire mass of growing culture 5° C or more per hour if the heat is not removed. Heat removal is slow because of the low heat transfer rate through the bran medium. The large bulk of material in a rotating drum may also result in a major problem of heat removal. In submerged culture, the same quantity of heat is liberated for the same carbohydrate consumption, but the heat is more readily removed because of better mixing and heat transfer in the fermenter, and greater dilution of medium (e.g. 1000 pounds solids in 10,000 pounds of medium).

There are several commercial enzymes which can be produced with the same efficiency by submerged or surface culture with respect to the units of activity produced per weight of nutrient consumed. The decision about which method to use in production is made on the basis of which one gives the fewest control problems. The selection of organism can also influence this decision. Frequently the choice is that process which gives the lowest level of contamination with undesirable enzymes. Sometimes the superiority of the submerged process – e.g. growth of *Bacilli* to produce liquefying diastase and alkaline protease – is overwhelming, and the decision is clear. There are, however, some fungal cultures which produce activity poorly in submerged culture – e.g. *Rhizopus* to produce amyloglucosidase – and the surface method is the one of choice.

c) Differences in Laboratory and Commercial Enzyme Processing

There are also major differences between large-scale enzyme processing and the usual laboratory methods of preparing enzymes. A complete laboratory procedure involving the cooling of the grown culture, extraction, filtration and precipitation can often be carried out in less than four hours with good temperature control at all steps. Drying generally adds another two hours. Normal commercial transfer (i.e. material handling) and refrigeration facilities usually require an 8 to 16 hour period for each required operation. The time scale, therefore, increases by a factor of about ten. Most enzymes are processed at

5–10° C at all stages up to the drying stage to minimize the effect of longer processing times, but microbial control requires close attention even at these temperatures. Preservatives may be added to the extracts during processing to minimize the growth of contaminants.

Drying varies greatly depending on the enzyme. Vacuum driers at low temperature may be required, but most commercial enzymes can be dried in a current of air at 40–50° C with only slight loss in activity. Some enzymes (e.g. laundry enzyme) are now being dried in spray driers.

Development of a new enzyme requires laboratory and plant work to determine the correct processing conditions. Modifications in processing may be required to increase the yield to a profitable level. Several industrial enzymes show a recovery of 70–80% of the culture activity in the final dry product. Liquid products show a higher recovery because the final precipitation and drying steps are eliminated.

3. Applications

Enzymes derived from animal and plant sources have been employed in relatively large amounts for many years. Many of these applications will probably continue to use the enzymes derived from these sources, because certain of these enzymes have specific qualifications for the purpose or involve an inexpensive byproduct from some other process. As the cost of the plant or animal source increases, the search for a microbial replacement will be accelerated. Some commercial applications of enzymes, especially those of microbial origin, represent new, rapid, controlled processes for carrying out conversions which were formerly brought about slowly or erratically by microorganisms contaminating a natural raw material. There are a number of industrial applications in which the living cell continues to be the source of microbial enzyme; the best known of these is the use of yeast in alcohol production.

Although a process for the production of diastatic enzyme from fungi was patented (Takamine, 1894) nearly seventy five years ago, the contamination of the preparation with bacteria prevented its use as a replacement for malt diastase. Almost 40 years elapsed before other microbial enzymes were developed and found their place in industrial processes. These microbial enzymes have either supplanted, partially replaced, or surpassed plant and animal enzymes.

Microbial enzymes are generally not exact counterparts of the plant or animal enzyme which they replace, but there has been an unfortunate tendency to carry over the nomenclature from the past and apply it to the microbial enzyme. This has been particularly true in the case of

Table 5. *Uses and sources of industrial enzymes*

Industry	Application	Enzyme	Source	US Suppliers[a] of microbial enzyme for use
Baking	Bread dough modification	protease	fungal	4, 7, 10
	Flour supplementation	amylase	malt at mill	—
			fungal at bakery	4, 7, 10
	Bleaching of natural flour pigments	lipoxidase	soybean flour	—
Beer (Brewing)	Mashing	amylase	malt	—
	Chillproofing	protease	fungal, bacterial, papain, pepsin	10
	Low dextrin (low calorie)	amyloglucosidase	fungal	4, 10
Candy	Soft center candies, invert sugar fondants	invertase	yeast	1, 8, 9, 10
(Confectionery)	Recovery of sugar scrap	amylase	bacterial	7
Cereal	Preparation of precooked cereals	amylase	malt, fungal	4, 7, 10
Coffee (Cocoa)	Removal of mucilage from berries	pectinase	fungal	7
	Thinning of concentrates	amylase, pectinase	fungal	4, 7, 10
Dairy	Milk curd formation for cheese making	rennet	fungal	5
			calf stomach	—
	Protein hydrolyzates	proteases	papain, pancreas, fungal, bacterial	4, 7, 10
	Removal of oxygen from egg whites	glucose oxioxidase + catalase	fungal	1, 4

[a] *Listing of United States Suppliers of Microbial Enzymes:*
1. Fermco Laboratories,
2. Grain Processing Corporation,
3. Lederle Laboratories,
4. Miles Laboratories, Inc.,
5. Chas. Pfizer and Company,
6. Premier Malt,
7. Rohm and Haas Company,
8. Standard Brands Inc. (Clinton Corn Processing Company),
9. Universal Foods,
10. Wallerstein Company.

Table 5. (continued)

Industry	Application	Enzyme	Source	US Suppliers[a] of microbial enzyme for use
Distilling	Premalt (liquefication)	amylase	bacterial malt	4 —
	Mash conversion (saccharification)	amylase amyloglucosidase	malt fungal	— 2, 4
Fishing	Thinning of condensed solubles	protease	bacterial	7
Fruits Fruit juices	Clarification, filtration and concentration of juices	pectinase	fungal	4, 7, 10
	Debittering of grape fruit	naringinase	fungal	7
Laundry aids and dry cleaning	Presoak and washing	alkaline protease	bacterial	5
	Spot removal	protease, lipase, amylase	mixture of fungal, bacterial, pancreatic	10
Leather	Bating	protease	pancreatic, bacterial fungal	7, 10
Meat packing	Tenderizers	protease	papain, bromelain fungal	7, 10
	Recovery of meat scraps	protease	bacterial	7
Oil	Fracturing of oil wells	hemicellulase-cellulase	fungal	7
Paper	Modification of starches for sizes and coatings	amylase	bacterial	4, 7, 10
Pharmaceutical	Digestive Aids	protease, lipase, amylase	pancreatic fungal	— 4
	Enhanced spreading of injectables	hyaluronidase	animal testis	—
	Test papers for glucose	glucose oxidase + peroxidase	fungal, horseradish bacterial, pancreatic, bromelain	4 3
	Antiinflammatories	Streptokinase, trypsin	bacterial, pancreatic	3

7*

Table 5. (continued)

Industry	Application	Enzyme	Source	US Suppliers[a] of microbial enzyme for use
Photographic	Recovery of silver from used films	amylase	fungal	4, 7, 10
Starch	Various corn sirups	amylase	fungal	4, 7, 10
	Dextrose production	amyloglucosidase	fungal	4, 2
	Fructose production	glucose isomerase	microbial and used internally	8
Textile	Starch desizing	amylase	bacterial	4, 6, 7, 10
Wines	Improved yield of juice Clarification and filtration	pectinase	fungal	4, 7, 10

microbial enzymes which replace those derived from malted cereals. Although the microbial enzyme may not be an exact counterpart of the enzyme derived from other sources, it is usually possible, by selection of organism and growing process, to tailor-make an enzyme which can perform under the required conditions. Some times combinations of microbial enzymes are required.

It may be well to review the progress which has been made in the replacement and supplementation of commercial plant and animal enzymes during the period from 1930–1969 and also those commercial processes which have evolved from applications of microbial enzymes. A very comprehensive monograph covering the use of enzymes in food is available (Reed, 1966). A fairly complete listing of enzymes used commercially, and of their sources is given in Table 5.

a) Commercial Enzymes of Plant Origin and the Extent of their Replacement by Microbial Enzymes

α) Cereal Amylases

Cereal amylases represent the most widely used commercial enzymes. The amylases derived from malted barley, and to a lesser extent malted wheat, have been employed for centuries in the preparation of malted beverages. In this application, both the liquefying endoamylase (α-amylase) and the saccharifying exoamylase (β-amylase) are required to bring about the required liquefaction of the grain in the mashing process, and the production of fermentable sugar (maltose) utilized by the yeast to produce alcohol. The liquefaction of the starch in the mash can be carried out by the microbial liquefying amylase obtained from *Bacillus subtilis*. The liquefying amylase derived from Aspergilli (flavus-oryzae generally) is less suitable because of its lower heat stability. The saccharification of the starch may be accomplished by amyloglucosidase derived from either *Aspergillus niger* or *Rhizopus oryzae*. Fungal amyloglucosidase produces glucose rather than maltose, but the extent of conversion to fermentable sugar is greater than with malt α-amylase, because less reversion products (i.e. isomaltose) are produced. The yields of alcohol are therefore improved by the fungal supplements. In addition, the amyloglucosidase is less affected than malt amylase by the presence of bacteria and low pH in the fermentation step. In the production of beverage alcohol the complete replacement of malt is not indicated, because the malt reportedly imparts certain desirable flavours not obtained with the fungal preparations. Large distillers are interested in preparing their own amyloglucosidase for captive use (van Lanen and Smith, 1969). Malt still occupies a strong position in the preparation of beer because of considerations of flavor. One commercial application of fungal amyloglucosidase in the preparation of low dextrin beer has been disclosed (Gablinger, 1968), and has had wide promotion as "low calorie beer". The resulting beer also has improved clarity.

The cereal amylases also find extensive use in the baking field as flour supplements. It is the α-amylase activity which must be added to flour to give the needed high gassing power (active release of CO_2) in fermenting dough. Flour usually has an excess of β-amylase, but, since the early part of the 20th century, malt α-amylase in the form of malted wheat or barley flour has been added at a level of 0.25–0.5 % at the mill to bring about nearly complete conversion of damaged starch to fermentable sugar. In modern bakery practice fungal α-amylase (from *Aspergillus flavus oryzae*) is added to increase the fermentation rate at the dough stage and impart other desirable properties, such as crust color and soft crumb, to the finished bread. The fungal supplements are more suitable than additional malt, because the fungal supplements can be prepared at high activities, either without protease or with a predetermined level of protease. The fungal supplements also have a lower level of bacterial contamination. The capability of controlling the protease level in the supplement is critically important in baking operations. The ease with which this factor can be controlled by standardized fungal preparations has stimulated this application of microbial enzymes. Fungal protease in carefully measured amounts are used in commercial baking operations to lower the time and work input required to mix and mellow dough prepared from modern strong flours. Tabletted forms of diastase and protease provide a wide latitude in the baker's choice of flour supplements.

A number of other commercial processes involving starch modification which formerly employed malt amylase, such as the desizing of starched textiles and the preparation of starch sizes and cold swelling starches, have been captured by microbial enzymes. These applications generally use amylases derived from *Bacillus subtilis*, because the high thermal stability of these enzyme preparations allow operation at high temperature and because the less extensive hydrolysis of the starch by the bacterial preparations gives more desirable products. The liquefying activity of bacterial amylases can usually be prepared more cheaply than the fungal amylase.

β) Plant Proteases

Three other plant enzymes, papain, bromelain, and to a lesser extent ficin, have found acceptance in the food industry as proteases. Papain is derived from the latex of the fruit, leaves, and trunk of *Carica papaya*, and bromelain from the fruit and stems of pineapple plants. These enzymes are used to prevent the hazing of beer when chilled (Chill-Proofing) by modifying the protein. Other applications for these plant proteases are in meat tenderizers and digestive aids. Ficin from the latex of *Ficus carica* is used to a much lower extent, perhaps because of its marked action on native protein and difficult handling. Proteases from *Aspergillus flavus-oryzae*, and to a lesser extent from *Bacillus subtilis*, have been used to replace and supplement these plant proteases in all applications, but papain continues to have the widest acceptance.

γ) Lipoxidase

Lipoxidase derived from soybean flour is widely used to decolorize the natural pigments in wheat flour. Although lipoxidase activity is exhibited by many microbial (fungal) preparations, these have not been able to replace the soybean lipoxidase.

b) Commercial Enzymes of Animal Origin and the Extent of their Replacement by Microbial Enzymes

α) Pancreatic Preparations

Crude and purified proteases derived from the pancreas have a long history of commercial use. Many of the applications employ a cruder form, pancreatin, but some drug applications require purified trypsin.

Processes carried out at a pH of 7–9 represent the most efficient use of this type of enzyme. The first industrial process involving this type of protease was in the bating of hides to remove the debris from the skin after liming, and to impart softness and air exchange to the finished leather. Although pancreatic bates are still in general use proteases derived from *Aspergillus flavus-oryzae* and *Bacillus subtilis* now supplement them.

Pancreatic preparations are also used in the preparation of hydrolyzed protein products and bacteriological media. Fungal and bacterial proteases alone and in mixture can be used to prepare similar products. The cost and attainment of needed properties determine the choice of enzyme.

Pancreatic preparations have been widely used as digestive aids, because they contain proteases, amylase and lipase. They have been prescribed for patients who have pancreatic disorders or after removal of the pancreas. The various activities present in the pancreatic preparations can be duplicated by *in vitro* methods from blends of microbial enzymes derived from *Bacillus subtilis*, *Aspergillus flavus-oryzae* and *Aspergillus niger*. Cellulase derived from Aspergillus niger is often added to the microbial preparation. The pancreatic preparations still hold the major share of the market, but this could be a useful application for the right combination of microbial enzymes.

Pancreatic preparations also make up a large part of the "spot removal" enzymes employed in dry cleaning establishments to remove stubborn stains. The activity of these preparations is usually enhanced by adding microbial amylase, protease, and lipase. Pancreatic preparations were patented (Röhm, 1913) and used in Europe as presoaks long before the era of modern-day laundry aids. These applications represent the forerunner of enzyme-containing detergents, but the pancreatic preparations could not withstand the high temperatures and pH used in the washing cycle. The development of alkaline protease from selected strains of *B. subtilis* has solved many of the problems associated with this application, and advertising has accomplished the rest of this development.

β) Pepsin

Pepsin derived from the mucosa of hogs has found use in chillproofing beer and as a digestive aid. Microbial proteases (and papain) can replace pepsin in chillproofing, but there are no commercially available microbial proteases which show the low pH optimum (1.8–2.2) exhibited by pepsin.

γ) Rennet

Rennet is used extensively in the production of cheese. This enzyme is found in the fourth stomach of the calf; it converts casein into para-casein, which, in the presence of calcium, precipitates to form an elastic curd. Much commercial rennet is contaminated with pepsin which tends to act strongly on casein and thereby produces a weak curd with off flavor. Many microbial proteases can clot milk, but only recently have microbial preparations been produced which can replace calf rennet to prepare cheeses of good flavor. Microbial preparations from *Endothia parasitica* and *Mucor pusillus* var. Lindt seem to be the most promising, and these have been used commercially. The new sources of rennet were reviewed recently (Sardinas, 1969).

c) Commercial Processes which Use Microbial Enzymes

α) Invertase

Invertase prepared from *Saccharomyces cerevisiae* and *S. carlsbergensis* is used in the preparation of soft center candies, fondants, and invert sugars.

β) Pectinases

Pectinases are prepared from various species of *Aspergilli* and *Penicillia*. The commercial preparations contain a variety of pectic activities which include pectin esterase, polygalacturonase, polymethylgalacturonase and pectin transeliminase. Before these enzymes were introduced in the early 30's (Willaman and Kertesz, 1931), processors of fruit juices found difficulty in obtaining high yields, and high pressing or filtration rates. Shelf stability was also a problem. The pectinases are now used to aid in the clarification, filtration, and concentration of many fruit juices, especially apple, grape, and berry juices. The pectinases are valuable in improving the yield of juice in the preparation of wines and hasten the aging process. Concentrated fruit juices can be prepared, stored under refrigeration, and shipped without gelation. The juice concentrates can be diluted to single strength at any convenient location or time. Jellies may be prepared from the concentrates by adding pectin.

Pectinases are also used in the processing of green coffee beans to hasten the removal of the jelly which surrounds the coffee cherry. Natural fermentation may give a coffee bean of inferior quality.

γ) Glucose Oxidase

Glucose oxidase is produced by a number of *Aspergilli* and *Penicillia* (*A. niger* and *P. notatum*). It catalyzes the following reaction:

$$\text{Glucose} + O_2 + H_2O \xrightarrow{\frac{\text{glucose}}{\text{oxidase}}} \text{Gluconic Acid} + H_2O_2 \, .$$

It is used widely in specific test procedures for the presence of glucose, for example in the diagnosis of diabetes or in analyzing food products where the formation of hydrogen peroxide denotes the presence of glucose. It also finds use in quantitative analytical procedures for glucose in which the glucose oxidase reaction is coupled with horse-radish peroxidase and an organic compound acting as a hydrogen donor.

The largest commercial uses of glucose oxidase require the presence of catalase to remove the peroxide formed in the glucose oxidase reaction:

$$2\,H_2O_2 \xrightarrow{\text{catalase}} 2\,H_2O + O_2\,.$$

This system is used to remove sugar present in low concentration in egg whites. The system may also be used to remove small amounts of oxygen from the head space of sealed containers by adding a packet containing glucose, oxidase, and catalase.

δ) Sirup Enzymes

Certain amylase preparations derived from *Aspergillus oryzae* can increase the extent of conversion of corn sirup to fermentable sugars without producing the unwanted crystallization characteristic of sirups made by the simple acid hydrolysis of starch (Dale and Langlois, 1940). A considerable amount of work by manufacturers of corn sirup and enzyme suppliers has been carried out after this original observation. As a result a whole series of corn sirups representing various degrees of conversion and combinations of dextrose, maltose, and oligo-saccharides were developed using various microbial enzymes alone and in mixture. More recently starch has been converted almost quanti-tatively to crystalline glucose by the enzyme process. This new process awaited the development of a commercial source of amyloglucosidase. Amyloglucosidases derived from *A. niger* and *R. oryzae* are used in the production of glucose, and *A. oryzae* Amylase alone or mixed with amyloglucosidase is used to prepare the various sirup compositions. Some sirup manufacturers have found it advantageous to prepare their own amyloglucosidase.

Although acid hydrolysis is most commonly used to prepare the starting sirup, the liquefying amylase from *B. subtilis* can be used to thin the initial starch paste without the need for acid resistant tanks. The process then becomes an all-enzyme process. The all-enzyme process produces a smaller amount of reversion products than the acid-enzyme process.

ε) Glucose Isomerase

After manufacturers developed sirups with properties useful in various applications, it became clear that increasing the sweetness of sirups by converting a large part of the glucose to fructose would be a good thing. A corn sirup as sweet as sucrose could be produced in this way. The major drawbacks to the use of alkali to accomplish the desired isomerization are the low degree of isomerization (33%), the large amount of decomposition products, and the salt formed in neutralizing the alkali. There has been a great deal of interest in microbial glucose isomerase since 1960, particularly in Japan where sucrose is in short supply. It was reported (Takasaki et al., 1969) that the commercial production of glucose-fructose sirups with the glucose isomerase of Streptomyces is practical. Some U.S. producers of corn sirup are also marketing sirups prepared with glucose isomerase.

η) Laundry Enzymes

It was noted above that the use of pancreatic enzymes in the washing of clothes was first patented in 1913, but the unstable nature of pancreatic preparations limited their use to presoak operations. The use of pancreatic enzymes in Europe to presoak clothes never died out completely. This vast market was activated by the development of alkaline protease in 1963–1964 by Novo Industries of Denmark and Royal Netherlands Fermentation Industries from the submerged culture of selected strains of Bacillus subtilis. The microbial enzyme therefore resulted in this major development. The value of the enzyme lies in its ability to solubilize stains caused by proteinaceous material. The alkaline protease is stable at high pH in the presence of phosphates and other ingredients of detergents. Proteolytic activity is also well retained up to 60° C. It seems likely that presence of other enzymes (e.g. lipase, amylase) with high tolerance to heat and detergent would be good supplements to the protease and would extend the range of stains which can be easily removed. Some products now on the market do contain added amylase, but most do not. The development of the commercial enzymes used in detergents has been reviewed in a number of articles (Hoogerheide, 1968; Koch, 1969; Wieg, 1969). The development of laundry enzymes has stimulated the entry of several new manufacturers into the industrial enzyme field.

ϑ) Lactase

Fluid milk contains about 5% of lactose. Lactose has low solubility and low sweetness, and a significant part of the world's population does not tolerate lactose in the diet. For these reasons the conversion of lactose to the component sugars, glucose and galactose would be worthwhile. Commercial lactases have been developed from lactose-fermenting yeasts and shown to have utility in the hydrolysis of lactose in a variety of milk products (Sampey and Neubeck, 1955). The market for these lactase preparations has not developed to any extent. There has also been some interest in using lactase to hydrolyze the lactose in the milk solids added to bread dough to get a more lively fermentation. This process also is not used commercially.

d) Outlook for Microbial Enzymes

Although the transition from animal and plant enzymes to microbial enzymes was relatively slow in "catching on", it seems likely that the present trend will continue. Commercial facilities for preparing enzymes by growing microorganisms are now available in a number of companies. The fullest possible utilization of these plants is economically sound. The excess or idle capacity available in plants for making antibiotics offers a source of production for the submerged growth of microorganisms suitable for enzyme production. The state of the art for making enzymes has advanced rapidly over the last two decades, and the problems of scale-up from the laboratory to the plant are now less formidable. The relative ease (i.e. compared to plants or animals) by which microorganisms can be screened, mutated, and otherwise developed for the production of a specific enzyme reduces the necessity for finding plants or animals as enzyme producers. A much greater variety of organisms is available, and the microorganisms are much more efficient.

There are obvious hazards in looking for enzymes in the realm of pathogenic organisms, and there are, of course, limitations by the U.S. government which restrict the utilization of enzymes derived from microorganisms in foods until they have been shown to be safe. As a result of these regulations, only a limited number of microorganisms were generally recognized as safe (GRAS) at the time (1958) when the regulations went into effect. The GRAS list included certain enzymes prepared in accordance with sound manufacturing practice and derived from *Saccharomyces cerevisiae* (invertase), *S. fragilis* (lactase), *Bacillus subtilis* (carbohydrase and protease), *Aspergillus flavus-oryzae* group (carbohydrase and protease) and *A. niger* (carbohydrase, cellulase, glucose, oxidase, catalase, pectinase, and lipase). Certain other microorganisms have been specifically cleared after presenting proof of the safe nature of the organisms and enzymes to the proper authorities. The same restrictions do not apply at present in many other applications, but good manufacturing practice demands that neither the producer nor the user of an enzyme product be exposed to health hazards.

There are several areas where microbial enzymes may have a good chance of commercial success.

α) Medical Uses

Reports (Chang *et al.*, 1967; Chang and Poznansky, 1968) about investigations on the application of encapsulated microbial enzymes to medical uses are available. By careful modification of the material, capsules can be obtained which are acceptable to body tissues, i.e., they produce neither clotting nor antibody formation. In this, they have an advantage over enzymes covalently linked to insoluble substrates. The application of encapsulated enzymes to a system in which, for one reason or another, a necessary enzyme is deficient has been investigated. The capsules may be inserted into the animal, or blood may be passed over a shunt system containing capsules, which is outside the body. Catalase deficiency in mice was overcome by encapsulated catalase and urea in the blood was removed with encapsulated urease. Should the products of enzymatic action be toxic, other capsules containing adsorbents may be simultaneously incorporated to lessen the toxicity. The method appears to have potential in the treatment of various enzyme-deficiency diseases.

One of the more hopeful prospects for a significant volume of pure enzyme is in the field of tumor therapy. L-*asparaginase* can induce remission in certain tumors in the mouse, rat, and dog, and suppress human leukemia by depleting an amino acid essential for neoplastic cells. This discovery has stimulated the search for L-*asparaginase*. Asparaginases are available from a number of microbial sources including fungi, yeast, and bacteria, but thus far, tumorinhibitory activity has been demonstrated only with the asparaginase from *Escherichia coli*, *Erwinia aroidae*, and *Serratia marcescens*. Asparaginase from *E. coli* has been prepared on a large scale.

From preliminary experimental evidence, collagenase derived from *Clostridium histolyticum* may prove to be useful as a collagenolytic agent for the treatment of ruptured vertebral discs, the removal of necrotic tissue following cryoprostratectomy, and the promotion of allograft survival in teeth. This enzyme may have some therapeutic utility in medicine and surgery where debridement is a part of the treatment. The major areas of possible use are in the treatment of severe burns and various dermal lesions such as decubiti, Ischemic ulcers, and other infective soft tissue lesions.

β) Dental Applications

There have been a considerable number of investigations into the use of dextranase derived from a variety of sources to eliminate plaque formation on human teeth. Many of these studies have proved inconclusive, and a significant amount of further work will be necessary to decide if dextranase can remove all plaque formed on the human teeth, and thereby significantly reduce the amount of dental caries occurring in human populations. It is conceivable that, once an enzyme or a combination of enzymes is found which will effectively perform this function, it will be included in formulations by commercial concerns now producing dental cleaning materials.

γ) Food Processing

As mentioned above, enzymes have found significant use in various applications in the processing of food materials. It is conceivable that a considerable amount of enzyme will be produced for use as processing aids to provide more efficient recoveries of the desired materials than are obtained through the less controllable chemical or mechanical processes. It is also conceivable that a significant number of enzymes will be discovered in the near future which will in effect do nothing more than offer over-all cost reduction in performing various steps in existing manufacturing processes.

The mushrooming demand for snack foods has stimulated an ever-increasing demand for new and different flavors to satisfy this market. It is conceivable that a significant number of enzymes will be isolated that will, either directly or indirectly, assist in flavor modifications for the production of new snack foods. It is likely that attempts will be made to provide enzymes for the selective modification of protein and carbohydrates by shortening their chain length. They might then impart new and different flavors or functionality to snack foods. The oils in which many deep fat-fried snack foods are prepared could be changed by a limited hydrolysis with selective lipases to impart an entirely new and different flavor. Work along these lines is progressing at a number of flavor and food companies.

It is extremely likely that enzymes will be used to aid in solving socio-economic-religious problems which arise during attempts to feed the undernourished populations with proteins obtained from exotic sources. These unusual food sources might include alfalfa or other plants, raw cellulose, soya beans, fish, and oil-grown microorganisms. Enzymes will probably find extensive use in the removal of certain materials now contained in existing food sources that render them unacceptable for human use. One example of this is the high level of undesirable RNA contained in single-cell protein derived from microbial growth on petroleum hydrocarbons.

δ) Waste Treatment

With the present interest in maintaining a clean environment increased emphasis will be placed on waste treatment, be it solid, liquid, or air. It is likely that enzymes from microorganisms can assist in the degradation and effective destruction of solid and liquid wastes. In this area producers will not be restricted to organisms on the GRAS list to obtain the necessary activity. The enzymes now used in some of these applications fall considerably short of the desired goal.

Enzymes from microbial sources offer unlimited potential for efficient use in a wide diversity of applications. The successful exploitation of their capabilities requires only a combined effort between three essential disciplines, engineering, chemistry and biochemistry.

References

Anon: Chem. Eng. 77, 44—46 (1970).
Beckhorn, E. J., Labbee, M., Underkofler, L. A.: Agr. Food Chem. 13, 30—34 (1965).
Butterworth, T. A., Wang, D. I. C., Sinsky, A. J.: ACS Abstract Micro. No. 5, September 1969.
Cayle, T.: U.S. Patent 3,075,886, Jan. 29, 1963.
Chang, T. M. S., Johnson, L. J., Ransome, O. J.: Can. J. Physiol. and Pharma. 45, 705—715 (1967).
— Poznansky, M. J.: Nature 218, 243—245 (1968).
Clayton, R. K., Smith, C.: Bioch. Biophys. Res. Comm. 3, 143 (1960).
Corman, J.: U.S. Patent 3,335,066, August 8, 1967.
Dale, J., Langlois, D.: U.S. Patent 2,301,609, 1940.
Demain, A. L.: Lloydia 31, 395—418 (1968).
Dulaney, E. L., Dulaney, D. D.: Trans. N.Y. Acad. Sci. 29, 782—799 (1967).
Ebata, S., Santomura, Y.: Agr. Biol. Chem. (Tokyo) 27, 478 (1963).
Edebo, L.: In: Fermentation Advances, Edit. D. Perlman. New York: Acad. Press 1969.
Egorov, N. S., Toropova, E. G., Ushakova, V. I., Mikharlova, T. N., Mironov, V. A.: Antibiotiki (USSR) 10, 678—684 (1965).
Gablinger, H.: U.S. Patent 3,379,534, April 4, 1968.
Garbutt, J. T.: U.S. Patent 3,318,782, May 9, 1967.
Gerhart, J. C., Holoubek, H.: J. Biol. Chem. 242, 2886 (1967).
Ghose, T. K., Kostick, J.: ACS Abstract Micro. No. 4, September 1969.
Goldstein, L.: In: Fermentation Advances, pp. 391—424, Edit. D. Perlman. New York: Acad. Press 1969.
Hillcoat, P. L., Blakley, R. L.: J. Biol. Chem. 241, 2995—3001 (1966).
Hoogerheide, J. C.: Fette, Seifen Anstrichmittel 70, (10), 743—748 (1968).
Horowitz, N. H., Fling, M., Macleod, H., Watanabe, Y.: Cold Spring Symposium 26, 233—245 (1961).
Huotari, F. I., Nelson, T. E., Smith, F., Kirkwood, S.: J. Biol. Chem. 243, 952—956 (1968).
Hurst, T. L., Turner, A. W.: U.S. Patent 3,047,471, July 31, 1962.
Inamine, E., Logo, B. D., Demain, A. L.: In: Fermentation Advances, Edit. D. Perlman. New York: Acad. Press 1969.
Inglett, G. E.: U.S. Patent 3,101,302, 1963.
Johnson, J. A., Miller, B. S.: Cereal Chem. 26, 371—383 (1949).
Kerr, R. W.: U.S. Patent 3,017,330, January 16, 1962.

Kertesz, Z. I.: N.Y. State Agr. Exp. Station (Geneva, N.Y.) Tech. Bulletin 589, (1930).
Koch, O.: Seifen, Öle, Fette, Wachse, September 17, 1969.
Langlois, D.: U.S. Patent 2,305,168, 1942.
Levisohn, S., Aronson, A. D.: J. Bacteriol. **93**, 1023—1030 (1967).
Lilly, M. D., Dunnill, P.: In: Fermentation Advances, Edit. D. Perlman. New York: Acad. Press 1969.
Mandels, Mary, Weber, J., Parizek, R.: in manuscript (1970).
Mehlitz, A.: Biochem. Zeit. **221**, 217—231 (1930).
Neubeck, C. E.: Unpublished reports (1970).
Novick, A., Horinchi, T.: Cold Spring Harbor Symp. **26**, 239—245 (1961).
Pardee, A. B.: In: Fermentation Advances, Edit. D. Perlman. New York: Acad. Press 1969.
Reed, G.: Enzymes in Food Processing. New York: Academic Press 1966.
Reese, E. T.: Can. J. Microbiol. **14**, 377—383 (1968).
— Lola, J. E., Parrish, F. W.: J. Bact. **100**, 1151—1154 (1969).
— Maguire, Anne: Appl. Microbiol. **17**, 242—245 (1969).
— Mandels, M.: J. A.C.S. **80**, 4625 (1958).
Rohm, O.: U.S. Patent 866,411, 1908.
— German Patent 283,923, 1913.
Sampey, J. J., Neubeck, C. E.: Food Engineering **27** (1), 68, 180, 183 (1955).
Sardinas, A.: Process Biochemistry **4**, 13—16, 21 (1969).
Silman, J. H., Katchalski, E.: Ann. Rev. Biochem. **35**, 873 (1966).
Sirotnak, F. M., Donati,. G. J., Hutchinson, D. J.: J. Biol. Chem. **239**, 4298 (1964).
Suzuki, H., Yamane, K., Nisizawa, K.: In: Cellulose and Its Applications, Adv. in Chem. Series 95, Edit. Hajny, G. and Reese, E. T., ACS, Washigton, D.C. (1969).
Takamine, J.: U.S. Patents 525,819-25, 1894.
— Industrial Eng. Chem. **6**, 824 (1914).
Takasaki, Y., Kosugi, Y., Kanbayashki, A.: In: Fermentation Advances, Edit. D. Perlman. New York: Acad. Press 1969.
Terui, G., Takano, M.: J. Fermentation Technology (Japan) **38**, 29—40 (1960).
Underkofler, L. A., Fulmer, E. I., Schoene, L.: Ind. Eng. Chem. **31**, 734—738 (1939).
— Severson, G. M., Goering, K. J., Christensen, L. M.: Cereal Chem. **24**, 1—22 (1947).
— In: Cellulose and Its Applications, Adv. in Chem. 95, Ed. Hajny, G. and Reese, E. T., ACS, Washington, D.C. (1970).
Van Lanen, J. M., Smith, M. B.: U.S. Patent 3,418,211, Dec. 12, 1968 (1969).
Wallerstein, L.: U.S. Patents 995,820; 995,823-6, 1911.
Wang, D. I. C., Humphrey, A. E.: Chem. Eng. December **15**, 108—120 (1969).
Wieg, A. J.: Process Biochemistry **4**, 30—34 (1969).
Willaman, J. J., Kertesz, Z. I.: N.Y. State Agr. Exp. Station (Geneva, N.Y.) Tech. Bulletin 178 (1931).

W. T. Faith † and C. E. Neubeck   E. T. Reese
Rohm and Haas Co.   U.S. Army Natick Laboratory
Philadelphia, Pa.   Natick, Mass.

CHAPTER 5

Overproduction of Microbial Metabolites and Enzymes Due to Alteration of Regulation

A. L. Demain

With 9 Figures

Contents

Introduction

From the viewpoint of the microorganism, the art of fermentation represents an inefficient and wasteful process. Microorganisms have evolved over the years, developing better and better mechanisms to prevent overproduction of their metabolites. Yet we microbiologists and bioengineers are dedicated to increasing the inefficiency of fermentation organisms as we continue to work toward the goal of almost complete conversion of nutrient into product with as little as possible going into the microbial protoplasm (except, of course, if we are in the single-cell protein business).

All microorganisms must possess regulatory (control) mechanisms in order to survive. Very efficient organisms are tightly controlled. In fermentation organisms, controls are less rigid but nevertheless present. Our genetic and environmental manipulations eliminate or bypass the residual control mechanisms and thus increase fermentation yields.

This chapter is an attempt to summarize the major regulatory mechanisms operating in the microbial cell and to highlight the effect of these mechanisms on the overproduction of both small primary and secondary metabolites and of enzymes. The principles underlying the main genetic and environmental manipulations used in the development of fermentation processes will be highlighted rather than the details of particular fermentations. By this presentation, it is hoped that the chapter can be used as a guide for the rapid development of future processes for new products regardless of the chemical or biological nature of the product or the type of microorganism employed.

1. Coordination of Microbial Metabolism

The coordination of metabolic reactions is an absolute necessity for life processes. A living cell placed in an environment containing starch, ammonia, and minerals must hydrolyze the carbon source to glucose, degrade it to 3-carbon compounds through the Embden-Meyerhof pathway (EMP) or the hexose monophosphate pathway (HMP), and feed these through the tricarboxylic acid (TCA) cycle to provide energy and intermediates; at least 20 enzymatic steps are involved. The intermediates must be converted to about 20 amino acids, 4 ribonucleotides, 4 deoxyribonucleotides, 10 or so vitamins, several fatty acids, etc. We can assume that each of these biosynthetic processes requires about 10 enzymes. These building blocks then have to be polymerized into 1000 proteins, three types of ribonucleic acid (RNA), deoxyribonucleic acid (DNA), mucopeptides, polysaccharides and lipids, again involving many steps. It is obvious that hundreds of enzymes must be formed and function in an integrated manner so that molecules are made in the correct proportions and valuable nutrients are not wasted. Only under such tight coordination can we expect cells to duplicate themselves in times as short as 20 minutes. Since a typical bacterial cell has the genetic potential to form over 1000 enzymes, coordination functions to insure that, at any particular moment, only necessary enzymes are made, that the correct amount of each are made and that once made, their activities are regulated by activation and inhibition. Thus, despite a constant genotype, microbes have an amazing ability to change their composition and metabolism in response to changes in their environment. The environment never alters the genome but markedly affects its phenotypic expression. The chief regulatory mechanisms which control the flexibility of microorganisms and tend to prevent oversynthesis of intermediates and end products are induction, catabolite regulation, and feedback regulation. In the following paragraphs and in Fig. 1, these three mechanisms are briefly described. The genetic mechanisms involved are beyond the scope of this chapter; the excellent review of Martin (1969) is highly recommended for their elucidation.

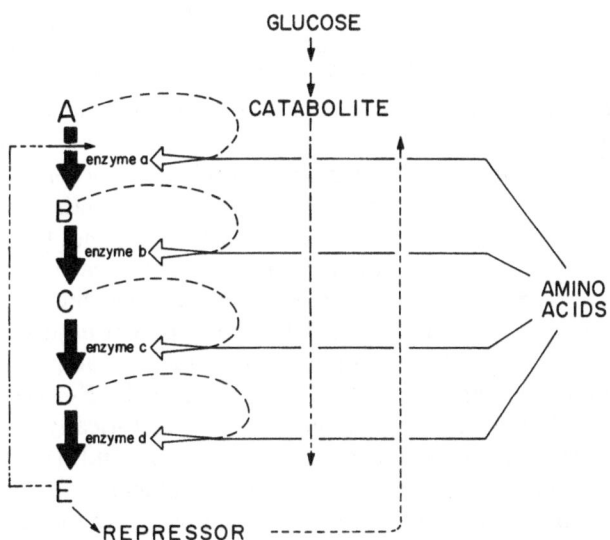

Fig. 1. Simple biochemical pathway subject to induction (———), feedback inhibition (— - - -), feedback repression, (- - - - -), and catabolite repression (— - — -)

a) Induction

Of the many enzymes that a cell can produce, a certain few, i.e. the "constitutive" enzymes, are always made in substantial concentration under all growth conditions. These include the enzymes of the HMP pathway, for example. Others, the "inducible" enzymes, are formed only when their substrates or substrate analogs are present in the medium. Substrate analogs which are not attacked by the enzyme ("gratuitous" inducers) are often excellent inducers. Inducible enzymes become necessary when the microorganism finds itself in the presence of a compound such as a polysaccharide, an oligosaccharide, or an amino acid as the sole source of carbon and energy. Thanks to the process of induction, enzymes are rapidly formed when needed, and energy and amino acids are not wasted on the making of unnecessary enzymes. Although most studies on induction involve the addition of inducer to the medium, induction can occur via the internal formation of inducer. For example, the inducer for pyruvate dehydrogenase is pyruvate which is formed by intracellular metabolism of glucose (Henning *et al.*, 1968).

b) Catabolite Regulation

The cell faces a problem when more than one utilizable growth substrate is present. Enzymes could be formed to catabolize all substrates, but this would be wasteful. Instead, enzymes are made which utilize the best substrate (usually glucose) and only after exhaustion of the primary substrate are enzymes formed which catabolize the poorer carbon source. The most common type of catabolite regulation, catabolite repression, involves the inhibition of formation of certain enzymes (i.e., repression) by the catabolic products of a readily utilizable carbon source. The repressed enzyme can be inducible or constitutive but in most cases, inducible enzymes are involved; in fact, very potent inducers can sometimes reverse catabolite repression. The classical example of catabolite repression is the repression of β-galactosidase in *Escherichia coli* by growth on glucose. Catabolite repression is historically known as the "glucose effect" but two facts should be kept in mind: (1) other rapidly used carbon sources can be as repressive as glucose; (2) in some cases, other carbon sources, such as acetate and citrate, are the repressors and glucose metabolism (uptake and catabolism) is the process which is repressed (Romano and Kornberg, 1968; Clarke and Lilly, 1969).

Under normal conditions, induction and catabolite repression together insure that inducible enzymes are only produced in the presence of substrate but that when several substrates are present, only the enzymes acting on the best substrate are formed. Very recent studies implicate inhibition of cyclic 3',5'-adenosine monophosphate formation as the key factor in catabolite repression (Perlman and Pastan, 1969). Cyclic 3',5'-adenosine monophosphate reverses catabolite repression of many enzymes in *E. coli* and its intracellular concentration is depressed 1000-fold by growth on glucose.

Two other types of catabolite regulation are known. In "catabolite inhibition", catabolites of a rapidly used carbon source inhibit the action of other enzymes. When an enzyme is actually inactivated by catabolites, the phenomenon is known as "catabolite inactivation". Much less is known about the significance of these forms of catabolite regulation.

c) Feedback Regulation

Whereas degradative enzymes are usually controlled by induction and catabolite regulation, the biosynthetic enzymes are chiefly controlled by feedback regulation. The two types included in this category are feedback (or end product) inhibition and feedback repression.

Feedback inhibition is the phenomenon by which the final metabolite of a pathway inhibits the action of an early enzyme of the pathway, usually the first enzyme. The end product inhibitor needs not resemble the substrate in size, charge, or shape, and binds to the enzyme at a site different from the substrate site. Occupancy of the "regulatory" site by the inhibitor distorts the enzyme molecule, interfering with binding at the substrate site – an "allosteric" effect. It is possible *in vitro* to treat the enzyme so as to eliminate susceptibility to feedback inhibition without destroying catalytic ability. In fact, aspartate transcarbamylase, the first enzyme of pyrimidine synthesis, can be separated into two subunits, i.e., the regulatory subunit possessing the inhibitor site and the catalytic subunit with the substrate-binding site. These subunits can then be recombined into a complex similar to the original enzyme.

Feedback repression is the inhibition of formation of one or more enzymes in a pathway by a derivative of the end product. In many (but not all) amino acid biosynthetic pathways, the amino acid end product must first combine with its transfer RNA (tRNA) before it can cause repression. Feedback repression is a widespread regulatory device especially for the synthesis of molecules intended for incorporation into macromolecules, e.g. amino acids, purines, and pyrimidines. Synthesis of vitamins also appears to be controlled by feedback repression, as well as by catabolite regulation (Birnbaum et al., 1967; Sasaki, 1965; Newell and Tucker, 1966; Wilson and Pardee, 1962; Papiska and Lichstein, 1968). Regulation of vitamin synthesis is important since only a small number (probably about 1000) of vitamin molecules are required per cell whereas many molecules of an average amino acid (probably 50 million) are required. An extremely wasteful case of vitamin overproduction would develop if enzymes for vitamin synthesis were produced at the same rate and were as active as the amino acid biosynthetic enzymes.

d) Branched Pathways

Most biosynthetic pathways are not simple routes but are branched, yielding more than one end product. Since the early part of the pathway is common to all end products, feedback regulation by one end product could interfere with formation of the other end products and starve the cell. To avoid this situation, various preventive measures have evolved in microorganisms. Among the most well known are isoenzymes, concerted feedback regulation, and cumulative feedback regulation. These are diagramatically represented in Fig. 2 and are briefly described below.

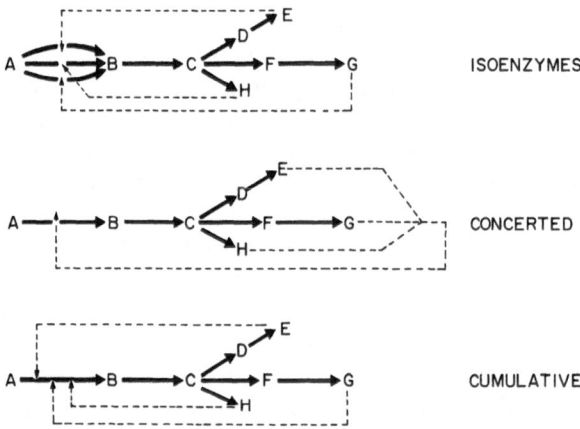

Fig. 2. Feedback regulation in branched pathways. Depicted are three of the protective devices used to prevent one end product from blocking synthesis of other end products. The regulatory effect (- - - - -) is applicable to both feedback inhibition and feedback repression

Isoenzymes. In this case, multiple enzymes are made; each carries out the same reaction but is regulated by a different end product. This mechanism is used in both feedback inhibition and feedback repression. A well known example of such control is the aspartic acid family in *E. coli* where the three aspartokinases are regulated by lysine, threonine and methionine respectively (Stadtman, 1968).

Concerted Feedback Regulation. This type is also known as "multi-valent" or "cooperative" feedback regulation. Only one enzyme is involved but more than one end product must be present in excess to inhibit or repress to a significant degree. This type of control is exerted in the branched-chain amino acid pathway of *Salmonella typhimurium* (Freundlich *et al.*, 1963).

Cumulative Feedback Regulation. In this case, each end product causes a partial inhibition or repression when present in excess alone but all end products must be present to effect complete blockage. The most well known example is the inhibition of glutamine synthetase in *E. coli* by eight end products (Stadtman, 1966).

In all three protective mechanisms mentioned above, the individual branches after the branchpoint act as simple linear paths being repressed and inhibited only by their own specific end product. Isoenzymes, concerted feedback and cumulative feedback are among the most common devices used to prevent an excess of one end product from drastically blocking the formation of others in branched pathways, but additional types have been discovered. The reader is referred to the paper of Datta (1969) for an excellent description of the entire field.

2. Overproduction of Metabolites

Due to some aberration in their regulatory mechanisms, organisms are often found in nature which overproduce metabolites. When these compounds have medical, nutritional, or industrial importance, the producing organisms are subjected to intensive development. Over a period of years, production of some fermentation products has been increased over 1000-fold by a combination of genetic and environmental manipulations. Environmental manipulations often involve the testing of hundreds of additives as possible precursors of the desired product. Occasionally, a precursor is found which increases production and/or directs the formation of one specific desirable product among a family of structurally similar products. Examples of such cases of "directed biosynthesis" include phenylacetic acid in the benzylpenicillin fermentation, amino acids in the production of actinomycins and tyrocidines, substituted benzoic acids in the formation of novobiocins and 5,6-dimethylbenzimidazole in the production of vitamin B_{12}. In many fermentations, however, precursors produce no benefits because their syntheses are not rate-limiting. In such cases, screening of additives has revealed dramatic effects, both stimulatory and inhibitory, of non-precursor molecules. These effects are due to interaction of non-precursor compounds with the residual regulatory mechanisms present in the fermentation organism.

a) Primary Metabolites

Primary metabolites are the vital small molecules of all living cells which are used as building blocks for essential macromolecules or are converted into coenzymes. The most important, from the industrial point of view, are the amino acids, nucleotides, and vitamins. The feedback type of regulation has proved to be most important in the fermentative biosynthesis of such products. In order to bypass feedback regulation, two types of manipulation have been employed: (1) decrease in the concentration of an inhibitory or repressive end product; (2) mutational alteration of the enzyme or enzyme-forming system to a condition less sensitive to feedback effects, i.e. mutation to feedback resistance.

Decrease in Concentration of End Products. In a simple pathway this principle is used to accumulate intermediates. For example, in Fig. 3, if one desires to accumulate compound C, an auxotrophic mutant is first

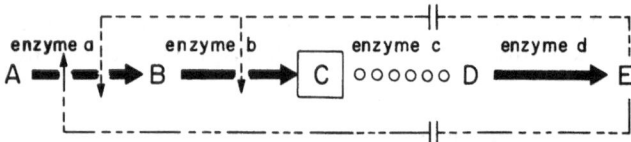

Fig. 3. Overproduction of an intermediate in a simple pathway. In parent organism, end product E feedback-inhibits (— - - -) the first enzyme and feedback-represses (- - - - -) both the first and second enzymes. A mutant is obtained which lacks (o o o) the third enzyme. E now must be supplied in the medium. If E is added at a growth-limiting concentration, feedback regulation is broken (⊣ ⊢) and C is overproduced (□)

obtained which lacks enzyme c. Such a mutant would require E for growth. By feeding only low levels of E, the organism cannot accumulate inhibitory or repressive levels of E in the cell. Thus, reactions a and b will operate unhindered by feedback regulation and high concentrations of C will be accumulated and excreted. By such a technique, an arginine auxotroph of *Bacillus subtilis* can accumulate 16 g/liter of citrulline (Okumura *et al.*, 1964) and arginineless *Corynebacterium glutamicum* produces 26 g/liter of ornithine (Kinoshita *et al.*, 1957).

Decreased end product concentration also yields valuable intermediates of branched pathways. Consider the pathway depicted in Fig. 4. End products L and N exert cumulative feedback repression and inhibition on enzyme a. In addition, L inhibits enzyme j_1 and N inhibits and represses j_2. By obtaining an auxotroph which lacks enzyme j_1, the organism now requires L for growth. By limiting the amount of L supplied from the medium, feedback effects of L are nullified. The organism now converts much of its carbon and nitrogen to compound J. Since j_2 is still regulated by N, only a little of J is converted on to N; most is excreted. The diagramatic scheme of Fig. 4 is actually a representation of the purine nucleotide pathway. [A = phosphoribosyl pyrophosphate (PRPP); J = inosine 5'-monophosphate (IMP); L = adenosine 5'-monophosphate (AMP); M = xanthosine 5'-monophosphate (XMP); N = guanosine 5'-monophosphate (GMP).] Thus, adenine auxotrophs of *C. glutamicum* and *Brevibacterium ammoniagenes* deficient in adenylosuccinate synthetase (enzyme j_1) accumulate as much as 13 g/liter IMP

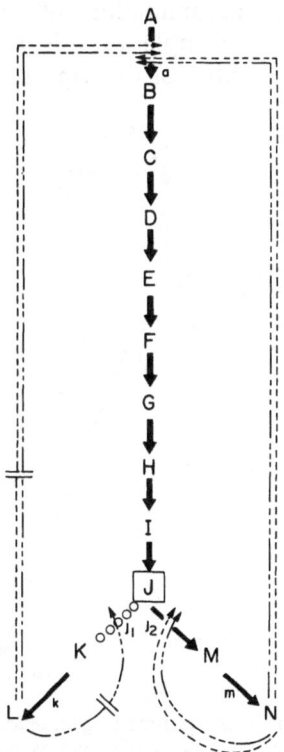

Fig. 4. Overproduction of an intermediate in a branched pathway. In parent organism, end product L feedback-inhibits (—- - - -) enzymes a and j_1 and feedback-represses (- - - - -) enzyme a; end product N feedback-inhibits and feedback-represses enzymes a and j_2. The feedback effects of L and N on enzyme a are of the cumulative type. A mutant is obtained which lacks (∘ ∘ ∘) enzyme j_1. L must now be supplied in the medium. If L is added at a growth-limiting concentration, feedback regulation by L is broken (⊣ ⊢) and J is overproduced (□). The scheme represents IMP overproduction by an adenineless mutant (see text)

(Nakayama *et al.*, 1964; Demain *et al.*, 1965; Furuya *et al.*, 1968). Inosine 5'-monophosphate has industrial importance as a flavor potentiator. Another flavor chemical (less potent than IMP) is XMP. It can be accumulated at levels of over 6 g/liter by removing XMP aminase (enzyme m) by mutation (Misawa *et al.*, 1964; Demain *et al.*, 1965;

Misawa *et al.*, 1969). Such cultures require guanine for growth and levels added to the medium must be kept low to bypass feedback effects on enzyme j_2 (IMP dehydrogenase) and enzyme a (PRPP amido-transferase).

If a pathway is branched, a decrease in the concentration of one end product often causes the accumulation of another end product. A commercially important example of this principle is the lysine fermentation. Lysine is a member of the aspartic acid family in which

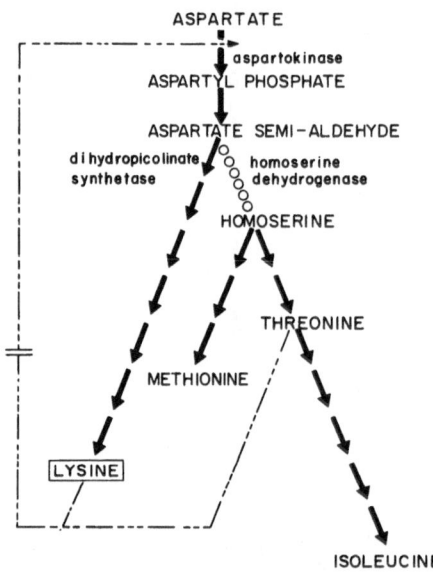

Fig. 5. The lysine fermentation. In the parent *C. glutamicum*, lysine plus threonine cause concerted feedback inhibition (— - - -) of aspartokinase. A mutant is obtained which lacks (○ ○ ○) homoserine dehydrogenase. Threonine plus methionine must now be supplied in the medium. If threonine is added at a growth-limiting concentration, concerted feedback inhibition is broken (⊣ ⊢) and lysine is overproduced. *C. glutamicum* lacks the usual feedback inhibition by lysine of dihydropicolinate synthetase

a branched pathway is used for biosynthesis of methionine, threonine and isoleucine in addition to lysine (Fig. 5). In *C. glutamicum*, and in *Brevibacterium flavum*, the single aspartokinase is controlled by concerted feedback inhibition requiring the simultaneous presence of excess lysine and threonine. By genetic removal of homoserine dehydrogenase, a mutant is obtained that requires both methionine and threonine to grow. Feeding low levels restricts the intracellular concentration of threonine thus bypassing the concerted inhibition of aspartokinase;

over 40 g/liter of lysine are excreted by the cultures (Nakayama *et al.*, 1966; Miyajima *et al.*, 1968). Another characteristic of the lysine over-producers is that the first enzyme of the lysine branch, dihydropicolinate synthetase, is insensitive to lysine inhibition. In *E. coli* and presumably in most organisms, this enzyme is inhibited by lysine. This regulatory idiosyncracy of *C. glutamicum* makes it an excellent organism for industrial lysine production.

In a fashion similar to the lysine fermentation, mutation of *E. coli* to lysine and methionine auxotrophy results in excretion of up to 4 g/liter of threonine (Huang, 1961), while isoleucine auxotrophy in *C. glutamicum* leads to 11 g/liter valine production (Nakayama *et al.*, 1961).

Mutation to Feedback Resistance. The easiest way to isolate either mutants that possess enzymes resistant to feedback inhibition or enzyme-forming systems resistant to feedback repression is to use toxic analogs (antimetabolites) of the desired compound. The plating of a population of cells on such an antimetabolite will kill most of the cells and select resistant mutants some of which overproduce the metabolite. The mechanism is as follows: an amino acid end product, for example, normally inhibits and represses its own biosynthetic enzymes and is also incorporated into protein. An antimetabolite also exerts inhibition and repression but cannot be used for protein synthesis. While the vast majority of the cells exposed to the antimetabolite die of amino acid starvation, those mutants which are insensitive to the analog can still make the amino acid and grow into colonies. Some of the mutants are resistant to the antimetabolite because of an alteration in the structure of the feedback-inhibited enzyme ("inhibition resistant") while others have an altered enzyme-forming system ("repression resistant"). Due to their altered controls, these mutants overproduce the amino acid. Table 1 lists the analogs which have been effectively used to obtain amino acid-, purine-, and pyrimidine-excreting mutants. Oversynthesis of the vitamins *p*-aminobenzoic acid, pyridoxine, and nicotinic acid has been obtained with mutants resistant to sulfonamide (Oakberg and Lauria, 1947), isoniazid (Scherr and Rafelson, 1962), and 3-acetylpyridine (Nakamura and Gowens, 1964) respectively. Mutations to inhibition resistance and to repression resistance, when genetically combined in a single strain, result in a marked synergy in overproduction. For example, when this was done with leucine excretors the resulting double mutant converted glucose to leucine in over 50% of theoretical yield (Calvo and Calvo, 1967).

The principle of mutation to feedback resistance was recently combined with that of reducing end product concentration in the development of an improved threonine fermentation. Shiio and Naka-mori (1969) isolated a threonine-excreting (1.9 g/liter) mutant of *E. coli*

Table 1. *Selection of metabolite-excreting mutants with antimetabolites*

Excreted compound	Antimetabolite used	References
Arginine	canavanine	Maas, 1961
Phenylalanine	p-fluorophenylalanine	Volkova *et al.*, 1965
Phenylalanine	thienylalanine	Jensen and Nasser, 1968
Tyrosine	p-fluorophenylalanine	Cohen and Adelberg, 1958
Tyrosine	thienylalanine	Adelberg, 1958
Tyrosine	D-tyrosine	Champney and Jensen, 1969
Tryptophan	5-methyltryptophan	Moyed, 1960; Lim and Mateles, 1964
Tryptophan	6-methyltryptophan	Lingens *et al.*, 1964
Valine	α-aminobutyrate	Kisumi *et al.*, 1969
Isoleucine	valine	Ramakrishnan and Adelberg, 1964
Leucine	trifluoroleucine	Calvo and Umbarger, 1964
Leucine	4-azaleucine	Jensen, 1969
Threonine	α-amino, β-hydroxy-valerate	Cohen and Patte, 1963
Methionine	ethionine	Adelberg, 1958; Musilkova and Fencl, 1964
Methionine	norleucine	Rowbury, 1965
Methionine	α-methylmethionine	Lawrence *et al.*, 1968
Methionine	L-methionine-DL-sulf-oximine	Jensen, 1969
Histidine	2-thiazolealanine	Moyed, 1961a; Sheppard, 1964
Histidine	1,2,3-triazole-3-alanine	Jensen, 1969
Proline	3,4-dehydroproline	Baich and Pierson, 1965
Adenine	2,6-diaminopurine	Kalle and Gots, 1962
Uracil	5-fluorouracil	Kaplan *et al.*, 1969

by selection in the presence of α-amino-β-hydroxyvalerate. [earlier, Cohen and Patte (1963) had shown that such a mutation desensitized the threonine-sensitive homoserine dehydrogenase resulting in overproduction of threonine.] The resistant strain was next mutated to isoleucine auxotrophy apparently by elimination of threonine deaminase; this resulted in production of 4.7 g/liter of threonine. The increase was evidently due to elimination of the path from threonine to isoleucine and, because of low internal isoleucine concentration, the bypassing of the concerted feedback repression normally exerted by threonine plus isoleucine on several of the enzyme of threonine biosynthesis. Finally, the antimetabolite-resistant and isoleucineless strain was mutated to loss of methionine synthetase. This methionine auxotroph produced over 6 g/liter of threonine, presumably because of bypassing the normal methionine regulation of the branched pathway.

An additional means of desensitizing enzymes to feedback inhibition involves removal of the sensitive enzyme by mutation and replacement by a second ("reversion") mutation. Often some of the resulting revertants excrete the end product of the pathway. Thus, removal and replacement of IMP dehydrogenase in *C. glutamicum* resulted in GMP accumulation (Demain et al., 1966) while removal and replacement of threonine deaminase in *Hydrogenomonas* caused excretion of isoleucine (Reh and Schlegel, 1969). Apparently these manipulations lead to an enzyme with an altered amino acid sequence which, although retaining catalytic activity, no longer binds the feedback inhibitor.

b) Secondary Metabolites

Secondary metabolites are molecules produced by a narrow spectrum of organisms; they have no general function in life processes although they may be important to the particular producing organism. They are usually produced as mixtures of members of a closely related chemical family, e.g., there are at least three neomycins, five mitomycins, 10 bacitracins, four tyrocidines, eight aflatoxins, 10 polymyxins, over 10 natural penicillins, and more than 20 actinomycins. The ability of an organism to produce secondary products is easily lost by mutation ("strain degeneration"). The subject of secondary metabolism has recently been analyzed by Weinberg (1970).

Derepression of Enzymes Forming Secondary Metabolites. Another characteristic of secondary metabolites is that they are usually not produced during the phase of rapid growth (trophophase) but are formed during a subsequent stage known as the idiophase (Bu'Lock, 1967). This phenomenon was first observed during the early development of the penicillin fermentation and has now been found to be a characteristic of many fermentations (Table 2). The factor that triggers

Table 2. *Some secondary products produced in idiophase*

Actinomycin	Hadacidin
Aflatoxin	6-Methylsalicylic acid
Bacitracin	Mycobacillin
Chlortetracycline	Novobiocin
Circulin	Patulin
Edeine	Penicillin
Ergot alkaloids	Polymyxin
Erythromycin	Prodigiosin
Gentisyl alcohol	Pyocyanin
Gentisylaldehyde	Tyrocidine
Gibberellic acid	Streptomycin
Gramicidin S	

secondary product formation at the end of trophophase is unknown. Perhaps it is an internal inducer which accumulates at high cell densities – or perhaps production is inhibited during trophophase because of catabolite regulation resulting from rapid sugar utilization. Whatever the mechanism is, there is no doubt that at the end of trophophase marked changes occur in the enzymatic composition of the cells and enzymes specifically related to formation of secondary products suddenly appear. This derepression of enzyme synthesis is clearly observed in the streptomycin fermentation. Here a key enzyme of streptidine biosynthesis, amidinotransferase, makes its appearance when the fermentation is 30 hours old and increases in specific activity for the next 10–30 hours (Walker and Hnilica, 1964). The importance of this enzyme to streptomycin synthesis is shown by the fact that it is present only in streptomycin-forming microorganisms. Since appearance of the enzyme can be prevented by chloramphenicol, *de novo* protein synthesis after trophophase is involved. Another enzyme, streptidine kinase, is also formed only during idiophase. In the penicillin fermentation, two enzymes have been observed to increase markedly after trophophase; one activates phenylacetic acid (the side chain of benzylpenicillin) while the other, penicillin acyltransferase, attaches the activated (coenzyme A ester) phenylacetate to the penicillin nucleus, 6-aminopenicillanic acid (Brunner *et al.*, 1968). Pruess and Johnson (1967) found a sudden increase in penicillin acyltransferase at 40 hours, a plateau in specific activity from 50 to 80 hours, and a decline thereafter. These activities correlated fairly well with changes in penicillin production rates. Furthermore, the enzyme was not found in fungi which do not produce penicillin and its level was highest in superior penicillin-producing strains. In the actinomycin fermentation, Katz (1967) has shown that phenoxazinone synthetase, the enzyme responsible for forming the chromophore of the antibiotic (actinocin), appears after growth just prior to the onset of antibiotic formation. The production of gramicidin *S* can be carried out by cell-free extracts of idiophase *Bacillus brevis*, but the enzyme is not found in trophophase cells (Spaeren *et al.*, 1967; Tomino *et al.*, 1967). Bacitracin formation by *Bacillus licheniformis* can be inhibited by inhibitors of nucleic acid and protein synthesis when added after the trophophase (Weinberg and Tonnis, 1967). In the patulin fermentation, Bu'Lock (1965) and Bu'Lock *et al.* (1965) have shown that the sequential idiophase appearance and disappearance in the medium of 6-methylsalicylic acid, gentisyl alcohol, gentisylaldehyde, and finally the antibiotic, patulin, is due to the process of sequential enzyme induction. If *p*-fluorophenylalanine is added during 6-methylsalicylic acid formation, 6-methylsalicylic acid accumulates instead of disappearing and none of the subsequent compounds is formed. Furthermore, idiophase mycelium of *Penicillium urticae*, but not trophophase mycelium, can convert added 6-methylsalicylic acid to patulin.

Catabolite Regulation. Catabolite regulation was actually observed in antibiotic fermentations years before the general significance of the phenomenon was appreciated. During the early days of penicillin development in the 1940s, it was found that the rapidly used glucose was an extremely poor substrate for penicillin production. Lactose, on the other hand, was only slowly used but supported excellent penicillin yields (Johnson, 1952). The classic Jarvis and Johnson chemically defined medium contains a mixture of glucose and lactose. In such a medium, the glucose is rapidly used during trophophase. Upon glucose exhaustion, *Penicillium chrysogenum* is derepressed for lactose utilization and the idiophase begins. During the slow utilization of lactose, the antibiotic is produced in the absence of growth. Lactose is not a specific precursor for penicillin synthesis; its value lies in slow utilization. Today slow glucose feeding has replaced lactose in the penicillin industry. Apparently limiting the concentration of glucose keeps catabolites at a low level. Glucose regulation of antibiotic synthesis can also be observed with resting idiophase mycelium (Demain, 1968). A similar situation is seen in the cephalosporin C fermentation where the chemically defined medium contains the rapidly used glucose for trophophase development and the slowly used sucrose for the idiophase. The gibberellic acid fermentation employs two carbon sources, glycerol and lactose, but these can be replaced by slow feeding of glucose (Darken *et al.*, 1959). Production of siomycin by washed mycelial suspensions of *Streptomyces sioyaensis* is inhibited by glucose or acetate (Kimura, 1967). Catabolite repression of phenoxazinone synthetase, an obligatory enzyme of actinomycin synthesis, has been demonstrated (Marshall *et al.*, 1968). Inhibition of antibiotic synthesis by glucose has been observed with violacein (DeMoss, 1967), mitomycin (Kirsch, 1967) and bacitracin (Weinberg, 1967). The formation of enterotoxin *B*, which might be considered as a high molecular weight secondary product, is repressed by glucose in *Staphylococcus aureus* (Morse and Mah, 1969). Catabolite regulation also plays a role in determining the ratio of closely related antibiotics found in fermentation broths (Inamine *et al.*, 1969). In the streptomycin fermentation, both streptomycin and mannosidostreptomycin are produced concurrently. Towards the end of the fermentation, when glucose is depleted, there is a sudden appearance of mannosido-streptomycinase (α-D-mannosidase) which converts the mannosido-streptomycin to streptomycin. Thus, if too much glucose is added, one ends up with an undesirable mixture of the two antibiotics. Induction is also important here since the enzyme is inducible as well as repressed by catabolites. The inducer is mannan which is usually supplied by addition of distiller's dried solubles to the medium.

Feedback Regulation. Feedback regulation also appears to play a role in secondary metabolism. It was shown many years ago that chloramphenicol inhibits its own production at concentrations nontoxic for growth of *Streptomyces venezuelae* (Legator and Gottlieb, 1953). Addition of 6-methylsalicylic acid to idiophase mycelium of *P. urticae* inhibits its own synthesis (Bu'Lock and Shepherd, 1968). Stimulation of carotenoid overproduction by *β*-ionone in *Phycomyces blakesleeanus* appears to be due to its ability to interfere with normal feedback inhibition (Reyes *et al.*, 1964). The inhibition of penicillin production by lysine (Demain, 1957) seems to be due to feedback regulation by the amino acid of a branched pathway (Fig. 6) leading to both lysine and

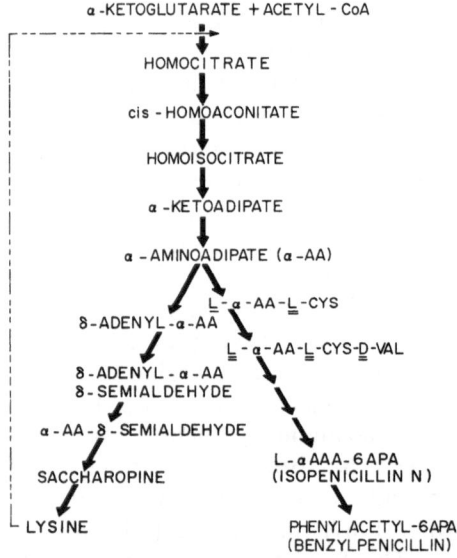

Fig. 6. Branched pathway leading to lysine and penicillin in *P. chrysogenum*. Feedback regulation shown as feedback inhibition (— - - -) of first enzyme but actual site and type of feedback regulation have not yet been established

penicillin (Demain, 1966). This hypothesis is supported by the experiments of Goulden and Chattaway (1968). These workers found that lysine auxotrophs blocked before α-aminoadipate (α-AA) required its addition for penicillin synthesis but lysineless mutants blocked after α-AA did not. Furthermore, in the latter type of auxotroph, accumulation of α-AA in the intracellular pool could only occur when the pool content of lysine was drastically reduced by starvation or growth.

Bypassing Control of Secondary Metabolism. Since the production of secondary metabolites is affected by genetically determined mechanisms − derepression (induction), catabolite regulation, and feedback regulation − it is clear that mutation should have a major effect on the production of secondary metabolites. Indeed, the chief factor responsible for the 100- to 1000-fold increases in production of antibiotics from the

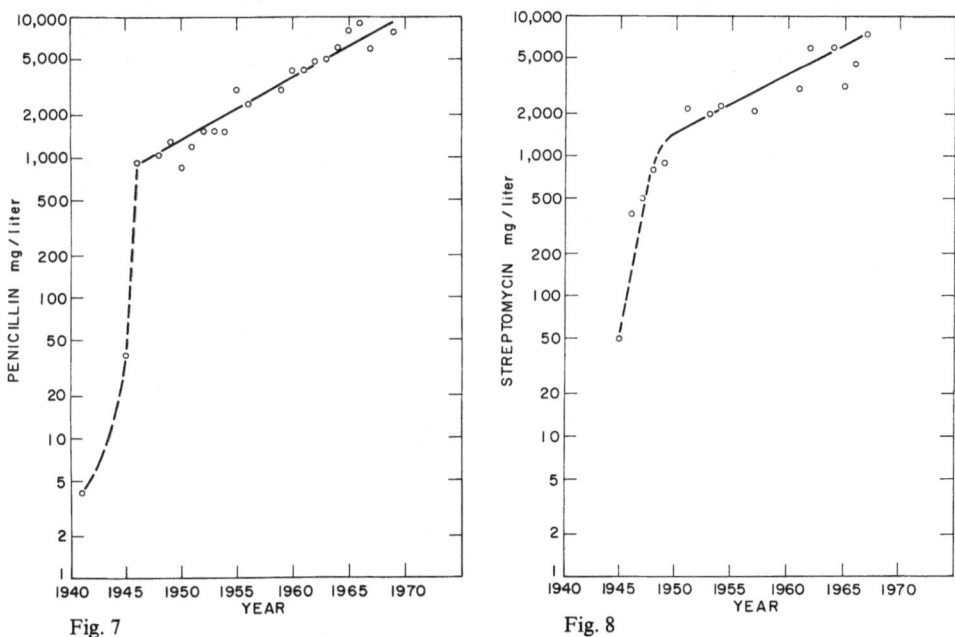

Fig. 7

Fig. 8

Fig. 7. Increase in penicillin broth potencies from 1941 to 1969. Figures were obtained from the literature, the points representing the highest published figure for each year

Fig. 8. Increase in streptomycin broth potencies from 1945 to 1967. Figures were obtained from the literature, the points representing the highest published figure for each year

time of their discovery to the present (Figs. 7 and 8) is mutation. Although most industrial strain improvement programs are based on mutation and random selection, several novel techniques designed to bypass regulation are worth mentioning. Use of the analog technique in the development of an antibiotic fermentation was demonstrated by Elander *et al.* (1967). Tryptophan (a precursor of the antibiotic, pyrrolnitrin) had to be added to the medium for optimal yields. By selecting for *Pseudomonas aureofaciens* mutants resistant to fluorotryptophan, Elander and his coworkers obtained a strain which produced three times as much antibiotic and no longer required supplementation with tryptophan. In those cases where an antibiotic is toxic to young cultures of the producing organism, the antibiotic itself can be used to select resistant cultures, some of which are higher producers. This has been done with streptomycin (Woodruff, 1966; Teteryatnic and Bryzgalova, 1968) and ristomycin (Trenina and Trutneva, 1966). The technique of mutation and reversion has been successfully used to obtain superior producers of chlortetracycline (Dulaney and Dulaney, 1967). Mutation to methionine auxotrophy and reversion increased yields threefold while mutation to nonproduction of the antibiotic followed by reversion resulted in a ninefold increase. The auxotrophic mutation in one branch of a pathway, increasing production by another branch, is probably the reason for the excretion of tetramethylpyrazine by an isoleucine-valine auxotroph of *C. glutamicum* (Fig. 9) (Demain *et al.*, 1967).

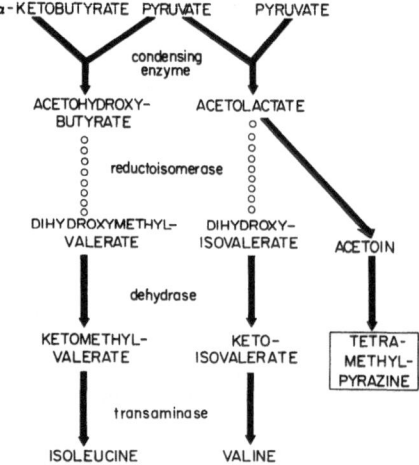

Fig. 9. Postulated pathway of tetramethylpyrazine formation in a mutant of *C. glutamicum* blocked in the isoleucine-valine pathway. The actual enzyme deficiency is not known but presumed to be the reductoisomerase

3. Overproduction of Enzymes

Enzymes have been used in the manufacturing industries, especially the food industry, and in medicine for many years. The properties of these proteins which lend themselves to extensive use include (1) rapidity and efficiency of action at low concentrations and under mild conditions of temperature and pH, (2) lack of toxicity, and (3) easy termination of their action by fairly mild treatment. The number of well discribed enzymes in the biochemical literature has markedly increased in recent years, from 80 in 1930 to 1300 in 1969.

Microbes have long been used as dependable sources of such industrial enzymes as amylase, invertase, lactase, protease, pectinase, catalase, and glucose oxidase. The interest in fermentation enzymes seems to be rapidly expanding as demonstrated by the current activity in the following areas:

Alkali- and temperature-resistant proteases (detergent industry),
Microbial rennin (cheese manufacturing),
Dextranase (inhibition of formation of dental caries),
Glucose isomerase (conversion of glucose to fructose),
Cellulase (conversion of cellulose to glucose),
Amyloglucosidase (conversion of starch to glucose),
Asparaginase (antitumor agent).

The amount of a particular enzyme produced by a particular micro-organism can vary tremendously since enzyme formation can be either stimulated or inhibited by control mechanisms. Catabolic enzymes have been known to vary several thousandfold, and hundredfold changes have been observed in the specific activity of biosynthetic enzymes (Clarke and Lilly, 1969). Since each control mechanism is influenced by environmental conditions, temperature, pH, medium composition, aeration, and stage of the growth are important in enzyme production. These conditions are usually determined empirically for each enzyme and each strain and will not be considered further unless they apply to a specific type of control mechanism under discussion. Many of the genetic and environmental manipulations quoted below were devised by virtue of detailed genetic knowledge of a particular species, but there is no reason to expect that they would not work on industrial micro-organisms. The excellent survey of Pardee (1969) forms the basis of the present discussion and additional examples are cited to bring the field up to date.

The manipulations to be discussed have been grouped into the following categories:

Environmental:
Addition of inducers,
Decrease in concentration of repressors;

Genetic:
Mutation to constitutivity and hyperproduction,
Increase in gene copies.

9*

a) Addition of Inducers

Some of the most potent inducers are nonmetabolizable substrate analogs ("gratuitous" inducers), e.g. isopropyl-β-D-thiogalactoside for β-galactosidase, methicillin for penicillin β-lactamase, malonic acid for maleate-fumarate cis-trans-isomerase and N-acetyl-acetamide for amidase. The effects of inducers are tremendous in some cases. Certain galactosides increase the specific activity of β-galactosidase in E. coli 1000-fold and result in cells containing several per cent of their total protein as this single enzyme.

If an inducer is particularly expensive or not readily available, it may be possible to substitute a compound which can be converted by the organism to the required inducer. For example, in Pseudomonas fluorescens, kynurenin induces kynureninase; the same effect can be obtained by adding tryptophan, a precursor of kynurenin, to the medium (Hayaishi and Stanier, 1951).

b) Decrease in Concentration of Repressors

The medium constituents are of utmost importance in production of enzymes repressible by end products or by catabolites of a rapidly used carbon source. The goal in such cases is a growth medium containing as little of the repressing compound as possible and one that does not lead to extensive internal formation of repressing compounds. Rich, complex media, as well as media containing rapidly utilizable sugars such as glucose, are to be avoided for maximum enzyme production.

Catabolic enzyme are usually under dual control of induction and repression: the repression can be caused by end products or by catabolites of rapidly used carbon sources or by both. For example, production of urease in Proteus rettgeri is induced by urea, repressed by ammonia, and is also affected by catabolite repression (Magana-Plaza and Ruiz-Herrera, 1967). Limitation of repressing compounds has a marked influence on enzyme levels. Inorganic phosphate, for example, represses phosphatase formation in E. coli and nucleases in Aspergillus quernicus. By limiting phosphate, the amount of alkaline phosphatase can be increased in E. coli from almost zero up to 5 % of the cell protein (Garen and Otsuji, 1964). Similar treatment of A. quernicus results in a 30-fold increase in 5'-nucleotide-forming nuclease, a 50-fold increase in the nuclease producing 3'-nucleotides, and a 40-fold increase in phosphatase (Ohta and Ueda, 1968). Repression of proteases by ammonia (Liu and Hsieh, 1969) and by amino acid mixtures in bacteria (Chaloupka and Kreckova, 1966a, b), and by sulfate in Aspergillus niger (Tomonaga, 1966) can be avoided by limitation of such compounds in growth media.

Many extracellular catabolic enzymes of industrial importance are subject to catabolite repression. Often high concentrations of inducers cause catabolite repression if they are used too rapidly. This has been found to be the case with cellulase, invertase, and dextranase (Reese *et al.*, 1969). The use of palmitate and acetate esters of inducers which are slowly hydrolyzed by the organism results in marked increases in enzyme production. For example, use of sucrose monopalmitate instead of sucrose increases invertase production 80-fold. Starch, maltose, glucose, and glycerol repress production of glucamylase by *Endomycopsis bispora* (Zickler, 1968). Hsu and Vaughn (1969) found that a pectic enzyme, polygalacturonic acid transeliminase, was repressed by its substrate and by all carbon sources readily utilized by *Aeromonas liquefaciens*; limitation of the carbon source by slow feeding of polygalacturonic acid increased production 1000-fold.

Repression of biosynthetic enzymes by end products can be prevented by limiting the production of these repressing compounds. This can be accomplished in two main ways.

Growth Limitation of Auxotrophic Mutants. Use of growth-limiting amounts of a required growth factor often derepresses enzyme synthesis. For example, leucine limitation in a leucine auxotroph of *S. typhimurium* yields a 40-fold increase in acetohydroxyacid synthetase (Freundlich and Trela, 1969). This same enzyme is increased fivefold when glycyl-L-valine, instead of valine, is fed to a valine auxotroph of *Hydrogenomonas* growing in a chemostat (Reh and Schlegel, 1969). A thiamine-requiring mutant grown with limiting thiamine shows derepressed levels of four enzymes involved in thiamine biosynthesis (Kawasaki *et al.*, 1969). The production of one of the enzymes, thiamine phosphate pyrophosphorylase, increases 1500-fold!

If the auxotroph has only a partial requirement, growth in minimal medium will be slow and derepression of the pathway enzymes will result. Thus, growth of a leaky pyrimidine auxotroph in minimal medium leads to a 500-fold increase in aspartate transcarbamylase (Moyed, 1961 b).

A variation on the above theme is the feeding of a slowly assimilated derivative or precursor of the growth factor to the auxotroph. For example, in *E. coli*, growth of a uracil auxotroph on dihydroorotic acid, instead of on uracil, gives derepressed levels of six enzymes of pyrimidine biosynthesis (Sheperdson and Pardee, 1960).

Analogs and Inhibitors. When we use prototrophic microorganisms, production of repressors can be reduced by addition of end product analogs to the medium. Thus, all 10 histidine pathway enzymes are derepressed up to 30-fold by 2-thiazolealanine (Ames and Hartman, 1963). In a similar manner, adenine, which inhibits thiamine synthesis, derepresses the enzymes of thiamine biosynthesis (Kawasaki *et al.*, 1969).

c) Mutations to Constitutivity and Hyperproduction

The aim here is to obtain "constitutive" mutants with altered regulatory genes so that the culture no longer requires an inducer or is no longer repressed by end products or by catabolites of a rapidly utilized carbon source. Thus, cultures are obtained which produce "induced levels" of enzyme without inducer or "derepressed levels" in the presence of compounds which normally cause enzyme repression. In a few cases, extremely high levels of enzymes are produced, much higher than the induced or derepressed levels of the parent strain. For example, aeration induces catalase in *Rhodopseudomonas spheroides* so that the enzyme level rises from 0.002 % to 1 % of the total extractable protein. Selection of mutants resistant to H_2O_2 results in cultures which produce catalase to the extent of 25 % of the cell protein (Clayton and Smith, 1960).

Removal of Inducer Requirement. a) Growth in chemostat with limiting substrate inducer. When growth of a population of an inducible parent culture is carried out in a chemostat with very low concentrations of a substrate inducer, selection occurs for constitutive mutants which do not require inducer. For example, growth in a chemostat with a low lactose concentration selects for *E. coli* mutants which form β-galactosidase without inducer and which contain as much as 25 % of their protein as β-galactosidase (Novick and Horiuchi, 1961).

b) Cycling with and without inducer. Successive growth cycles, alternating between a medium not containing inducer and one that does, enrich the population in constitutive mutants (Cohen-Bazire and Jolit, 1953). For example, the first growth cycle might contain glucose; both the rare constitutive mutants and the predominant β-galactosidase inducible parent cells will grow equally. Transfer of the mixed culture to a lactose medium will favor the constitutive mutant since the inducible parent must take time for induction. Repetition of these cycles of growth will finally allow the constitutive mutant to predominate since each transfer to glucose will return the inducible parent to the uninduced state.

c) Poorly inducing substrate. If the mutated population is grown in a medium containing as carbon source a compound which is a good substrate but a poor inducer, constitutive mutants are selected. An example is the use of phenyl-β-galactoside to select constitutive β-galactosidase cultures (Jacob and Monod, 1961).

d) Inhibitors of induction. There are compounds which block the induction of certain enzymes. Such compounds are cyanoacetamide in the case of *Pseudomonas aeruginosa* amidase (Clarke and Lilly, 1969) and 2-nitrophenyl-β-fucoside in the case of *E. coli* β-galactosidase (Cohen-Bazire and Jolit, 1953). In this procedure, cells are grown in a mixture of a substrate inducer and induction inhibitor. Since the induction of the enzyme is inhibited, only mutants which do not require inducer are able to grow.

e) Visual detection without enrichment. Methods have been devised to directly visualize a rare constitutive mutant colony among many inducible colonies on agar plates. Enrichment methods as described above are not required in these cases if mutation frequency is high enough. For example, mutation to β-galactosidase constitutivity can be detected by spreading a mutagenized culture on an agar medium containing glycerol as carbon source. After the colonies have grown, the plate is sprayed with o-nitrophenyl-β-D-galactoside. Only constitutive mutants have made β-galactosidase during growth in the absence of inducer. These colonies hydrolyze the colorless reagent releasing the yellow o-nitrophenol. The presence of one yellow constitutive mutant colony among hundreds of normal white inducible colonies is easily detected.

Resistance to End Product Repression. a) Antimetabolites. The use of toxic antimetabolites to select resistant cultures often yields mutants whose normal biosynthetic pathway enzymes are not repressed by the end product. For example, trifluoreoleucine selects mutants with derepressed levels (as much as 10-fold) of leucine biosynthetic enzymes (Calvo and Calvo, 1967). Certain canavanine-resistant mutants produce 30 times more of the arginine pathway enzymes than do their sensitive parents (Jacoby and Gorini, 1967). Mutants of *Lactobacillus casei* resistant to dichloroamethopterin are derepressed 80-fold in their ability to form thymidylate synthetase (Crusberg and Kisliuk, 1969). When *Diplococcus pneumoniae* mutants are selected on the basis of resistance to amethopterin, these cultures produce 100 times as much dihydrofolate reductase as the parent culture (Sirotnak et al., 1969).

b) Visual detection. Mutants not repressible by end products can be detected directly without enrichment. For example, mutation to phosphatase constitutivity can be detected after colonies have been grown in the presence of high phosphate and sprayed with p-nitrophenyl phosphate (Garen and Otsuji, 1964). Constitutive mutants are yellow, repressible colonies are white.

Another technique is that of spraying or overlaying colonies with a suspension of bacteria that require for growth a compound which is the end product of the pathway one desires to derepress. Constitutive mutants often overproduce the end product, excrete it, and cross-feed such tester cultures (O'Donovan and Gerhart, 1968).

Occasionally, constitutive mutants can be detected by their colonial appearance. For example, derepressed histidine mutants of S. *typhimurium* appear wrinkled in contrast to the smooth colonies formed by the repressible wild type (Roth and Hartman, 1965).

Pigment formation can be used to visually detect mutants constitutive for enzymes of the purine pathway (Dorfman, 1969). The
technique uses an auxotroph blocked early in the pathway which forms
red colonies because of the polymerization into a red pigment of
accumulated 5-amino-4-imidazole ribonucleotide. Normally, high levels
of adenine in the medium inhibit pigment formation because AMP
represses the pathway. After mutation of the auxotroph, the population
is plated on agar containing adenine; derepressed mutant colonies
are red.

c) Sporulation mutants. Protease formation, as well as sporulation
by bacilli, is inhibited by amino acid mixtures. Mutants which are able
to sporulate in the presence of an inhibitory mixture of amino acids are
easily selected by a heating procedure. Such mutants of *Bacillus cereus*
are derepressed for protease synthesis, producing about 10 times as
much protease as their parent (Levisohn and Aronson, 1967).

d) Reversion of auxotrophy. Mutation of a culture to auxotrophy
and reversion to prototrophy can be used to modify the regulatory
properties of the enzyme involved in the auxotrophic requirement. For
example, reversion of mutants lacking homoserine dehydrogenase in
one case resulted in a threefold increase in enzyme activity compared to
the original prototrophic culture (Patte *et al.*, 1963).

Resistance to Catabolite Repression. Mutants resistant to catabolite
repression are useful when one wants to obtain enzymes from cultures
grown in complex media or when nonrepressing carbon sources might
be too expensive for large scale fermentations. In addition, some mutants
resistant to catabolite repression produce extremely high levels of
enzymes. For example a yeast mutant, in which invertase production is
resistant to catabolite repression, produces almost 2 % of its cell protein
as invertase (Lampen *et al.*, 1967). Many of the mutants obtained so far
are apparently modified in their glucose catabolic pathway since the
production of many of their enzymes is no longer repressed by glucose.
However, there are some mutants in which resistance specifically applies
to catabolite repression of a single enzyme (Chasin and Magasanik, 1968).

a) Substrate of repressed enzyme as sole nitrogen source. Serial
transfer of *Aerobacter aerogenes* in a medium containing glucose plus
histidine (but no other nitrogen source) selects mutants resistant to
catabolite repression (Neidhardt, 1960). Normally histidine-degrading
enzymes are repressed by glucose, but in this case they are obligatory for
growth. The medium thus selects for mutants which produce histidase
in the presence of glucose. Such mutants show derepressed levels of
histidase, urocanase, and β-galactosidase when grown with glucose and
histidine. Growth of *P. aeruginosa* on succinate plus lactamide selects
amidase cultures resistant to catabolite repression. The wild-type

culture cannot grow since formation of the amidase is repressed by succinate, leaving no nitrogen available.

b) Cycling with and without glucose. *E. coli* shows a lag in growth when transferred from glucose to a medium containing lactose, maltose, acetate or succinate (Hsie and Rickenberg, 1967). This is evidently due to repression during growth on glucose of enzymes needed for utilization of these other carbon sources. Thus, alternation of growth between glucose and succinate, for example, selects mutants resistant to catabolite repression. The selection is also aided by the fact that the generation time in maltose, acetate, or succinate is markedly shorter for the mutant than for the repressible parent.

c) Visual detection. The methods described above for visual detection of other types of constitutive mutants can also be applied to mutants resistant to catabolite repression. For example, colonies are grown on agar containing glucose and are sprayed with *o*-nitrophenylgalactoside. If the parent is inducible, an inducer must be included in the agar; if the parent requires no inducer, it may be omitted. Mutants resistant to catabolite repression will be yellow. A similar technique was used recently to isolate mutants of *A. liquefaciens* resistant to catabolite repression in production of the extracellular enzyme, polygalacturonic acid transeliminase (Hsu and Vaughn, 1969). After colonies have grown on polygalacturonic acid one can detect the enzyme by flooding the plate with HCl. Clear zones indicate enzyme production. By replica-plating surviving colonies after nitrosoguanidine mutagenesis on polygalacturonic acid agar plates with and without glucose, one can detect repression-resistant mutants through their ability to produce clear zones on both types of plates.

d) Streptomycin dependency. For some unexplained reason, mutation to streptomycin dependency in *E. coli* decreases efficiency of glucose utilization, resulting in nonspecific resistance to catabolite repression (Couckell and Polglase, 1969).

d) Increase in Gene Copies

Occasionally some of the above techniques to select mutants producing high enzyme levels result in organisms with several copies of the gene dictating production of the enzyme (Novick and Horiuchi, 1961). Since it is generally true that enzyme productivity increases as copies of a gene increase, it would be desirable to intentionally introduce extra specific genes to a microorganism capable of producing a certain enzyme. This already has been done with *E. coli*, as will be described below.

Extrachromosomal DNA segments are present in many enterobacteria. These pieces of DNA, known as episomes (Novick, 1969), can be introduced into a culture by genetic means. Such a manipulation in *E. coli*, with an episome carrying the gene for β-galactosidase, increased enzyme production threefold (Jacob and Monod, 1961). A method is now available (Low, 1968) to obtain F' factors (episomes) containing any part of the *E. coli* chromosome, thus making it possible to increase production of a specific enzyme by episome transfer.

When a transducing phage carrying structural genes for bacterial enzymes is used to lysogenize an *E. coli* strain, and the prophage is subsequently induced to replicate (e.g. by UV treatment), the replication of the transducing phage results in an enormous increase in the number of gene copies per cell. At the same time, enzymes coded by these genes are synthesized at rates many times higher than normal (Buttin, 1963; Yarmolinsky, 1963). Since a method is now available that isolates specialized transducing phages for any *E. coli* gene (Gottesman and Beckwith, 1969), this technique of "phage-escape synthesis" should be extremely useful in increasing enzyme production.

4. The Future of Fermentation

The importance of regulatory mechanisms such as induction, catabolite regulation, and feedback regulation in the fermentative production of primary and secondary small metabolites and of enzymes has been described in these pages. Not included, but equally important, are the regulatory mechanisms affecting nucleic acid production. No longer can the fermentation microbiologist restrict his approaches to the empirical development procedures of the past. The future will see great attention in developmental research paid to the biochemical and genetic relationships between the microorganism and the environment in which it is placed. The bioengineer will be devoting more of his time to the designing of equipment that will monitor and control the level of inducers, repressors, and inhibitors added to the fermentation medium and produced by the organism. The fermentation microbiologist will become more and more influenced by the remarkable advances made in the genetics of microorganisms. The ease with which genes can be mapped in certain microorganisms and the recent demonstration of the ability to specifically mutate target genes in synchronized cells will be particularly attractive to developers of fermentations. We can expect to see mutants of well mapped organisms like *E. coli* and *B. subtilis* used for production of fermentation products, as well as for the acquisition of knowledge about the genetics of today's fermentation organisms. As the findings of the molecular biologists become successfully applied to medical problems, we can expect to see the emphasis of large-scale fermentation shift from small molecules to the production of complicated macromolecules. New techniques will not only be needed for the fermentation proper but, even more important, for the isolation of large molecules such as specific nucleic acids and enzymes. The fermentation field will no doubt change in the near future but it will be a change for the better – a change toward a more understandable, more exciting and more challenging level of scientific activity.

Acknowledgment

The preparation of this chapter was supported by Public Health Service Research Grant AI-09345 from the National Institute of Allergy and Infectious Diseases.

References

Adelberg, E. A.: J. Bacteriol. **76**, 326 (1958).
Ames, B. N., Hartman, P. E.: Cold Spring Harbor Symp. Quant. Biol. **28**, 349 (1963).
Baich, A., Pierson, D. J.: Biochim. Biophys. Acta **104**, 397 (1965).
Birnbaum, J., Pai, C. H., Lichstein, H. C.: J. Bacteriol. **94**, 1846 (1967).
Brunner, R., Roehr, M., Zinner, M.: Z. Physiol. Chem. **349**, 95 (1968).
Bu'Lock, J. D.: In: Biogenesis of Antibiotic Substances. Vanek, Z. and Hostalek (Eds.), p. 61. Prague: Publ. House Czech. Acad. Sci. 1965.
— Essays in Biosynthesis and Microbial Development. New York: John Wiley and Sons, Inc. 1967.
— Hamilton, D., Hulme, M. A., Powell, A. J., Smalley, H. M., Shepherd, D., Smith, G. N.: Can. J. Microbiol. **11**, 765 (1965).
— Shepherd, D.: Biochem. J. **106**, 29 p. (1968).
Buttin, G.: J. Mol. Biol. **7**, 610 (1963).
Calvo, J. M., Umbarger, H. E.: Federation Proc. **23**, 377 (1964).
Calvo, R. A., Calvo, J. M.: Science **156**, 1107 (1967).
Chaloupka, J., Kreckova, P.: Folia Microbiol. Prague **11**, 82 (1966a).
— — Folia Microbiol. Prague **11**, 89 (1966b).
Champney, W. S., Jensen, R. A.: J. Bacteriol. **98**, 205 (1969).
Chasin, L. A., Magasanik, B.: J. Biol. Chem. **243**, 5165 (1968).
Clarke, P. H., Lilly, M. D.: Symp. Soc. Gen. Microbiol. **19**, 113 (1969).
Clayton, R. K., Smith, C.: Biochem. Biophys. Res. Commun. **3**, 143 (1960).
Cohen, G. N., Adelberg, E. A.: J. Bacteriol. **76**, 328 (1958).
— Patte, J. C.: Cold Spring Harbor Symp. Quant. Biol. **28**, 513 (1963).
Cohen-Bazire, G., Jolit, M.: Ann. Inst. Pasteur **84**, 937 (1953).
Couckell, M. B., Polglase, W. J.: Biochem. J. **111**, 279 (1969).
Crusberg, T. C., Kisliuk, R. L.: Federation Proc. **28**, 473 (1969).
Darken, M. A., Jensen, A. L., Shu, P.: Appl. Microbiol. **7**, 301 (1959).
Datta, P.: Science **165**, 556 (1969).
Demain, A. L.: Arch. Biochem. Biophys. **67**, 244 (1957).
— Advan. Appl. Microbiol. **8**, 1 (1966).
— Lloydia **31**, 395 (1968).
— Jackson, M., Vitali, R. A., Hendlin, D., Jacob, T. A.: Appl. Microbiol. **13**, 757 (1965).

Demain, A. L., Jackson, M., Vitali, R. A., Hendlin, D., Jacob, T. A.: Appl. Microbiol. **14**, 821 (1966).
— — Trenner, N. R.: J. Bacteriol. **94**, 323 (1967).
DeMoss, R. D.: In: Antibiotics. Vol. II, Biosynthesis. Gottlieb, D. and Shaw, P. D. (Eds.), p. 77. New York: Springer-Verlag 1967.
Dorfman, B.: Genetics **61**, 377 (1969).
Dulaney, E. L., Dulaney, D. D.: Trans. N.Y. Acad. Sci. **29**, 782 (1967).
Elander, R. P., Hamill, R. L., Gorman, M., Mabe, J.: Abstracts of Papers, 154th Am. Chem. Soc. Meeting, Chicago, September, 1967, Q 42.
Freundlich, M., Burns, O., Umbarger, H. I.: In: Informational Macromolecules. Vogel, H. J., Bryson, V., and Lampen, J. O., Jr. (Eds.), p. 287. New York: Academic Press, Inc. 1963.
— Trela, J. M.: J. Bacteriol. **99**, 101 (1969).
Furuya, A., Abe, S., Kinoshita, S.: Appl. Microbiol. **16**, 981 (1968).
Garen, A., Otsuji, N.: J. Mol. Biol. **8**, 841 (1964).
Gottesman, S., Beckwith, J. R.: J. Mol. Biol. **44**, 117 (1969).
Goulden, S. A., Chattaway, F. W.: Biochem. J. **110**, 55 p. (1968).
Hayaishi, O., Stanier, R. Y.: J. Bacteriol. **62**, 691 (1951).
Henning, U., Dietrich, J., Murray, K. N., Deppe, G.: In: Molecular Genetics. Wittmann, H. G. and Schuster, H. (Eds.), p. 223. New York: Springer-Verlag 1968.
Hsie, A. W., Rickenberg, H. V.: Biochem. Biophys. Res. Commun. **29**, 303 (1967).
Hsu, E. J., Vaughn, R. H.: J. Bacteriol. **98**, 172 (1969).
Huang, H. T.: Appl. Microbiol. **9**, 419 (1961).
Inamine, E., Lago, B. D., Demain, A. L.: In: Fermentation Advances. Perlman, D. (Ed.), p. 199. New York: Academic Press, Inc. 1969.
Jacob, F., Monod, J.: J. Mol. Biol. **3**, 318 (1961).
Jacoby, G. A., Gorini, L.: J. Mol. Biol. **24**, 41 (1967).
Jensen, R. A.: J. Biol. Chem. **244**, 2816 (1969).
— Nasser, D. S.: J. Bacteriol. **95**, 188 (1968).
Johnson, M. J.: Bull. World Health Organ. **6**, 99 (1952).
Kalle, G. P., Gots, J. S.: Proc. Soc. Exp. Biol. Med. **109**, 277 (1962).
Kaplan, J. G., Lacroute, F., Messmer, I.: Arch. Biochem. Biophys. **129**, 539 (1969).
Katz, E.: In: Antibiotics. Vol. II, Biosynthesis. Gottlieb, D. and Shaw, P. D. (Eds.), p. 276. New York: Springer-Verlag 1967.
Kawasaki, T., Iwashima, A., Nose, Y.: J. Biochem. Tokyo **65**, 407 (1969).
Kimura, A.: Agr. Biol. Chem. Tokyo **31**, 845 (1967).
Kinoshita, S., Nakayama, K., Udaka, S.: J. Gen. Appl. Microbiol. Tokyo **3**, 276 (1957).
Kirsch, E. J.: In: Antibiotics, Vol. II, Biosynthesis. Gottlieb, D. and Shaw, P. D. (Eds.), p. 66. New York: Springer-Verlag 1967.
Kisumi, M., Komatsubara, S., Chibata, I.: Amino Acid Nucleic Acid Tokyo **19**, 1 (1969).
Lampen, J. O., Neumann, N. P., Gascon, S., Montenecourt, B. S.: In: Organizational Biosynthesis. Vogel, H. J., Lampen, J. O., and Bryson, V. (Eds.), p. 363. New York: Academic Press, Inc. 1967.
Lawrence, D. A., Smith, D. A., Rowbury, R. J.: Genetics **58**, 473 (1968).
Legator, M., Gottlieb, D.: Antibiot. Chemotherapy **3**, 809 (1953).
Levisohn, S., Aronson, A. I.: J. Bacteriol. **93**, 1023 (1967).
Lim, P. G., Mateles, R. I.: J. Bacteriol. **87**, 1051 (1964).
Lin, P. V., Hsieh, H. C.: J. Bacteriol. **99**, 406 (1969).
Lingens, F., Kraus, H., Lingens, S.: Z. Physiol. Chem. **339**, 1 (1964).

Low, B.: Proc. Nat. Acad. Sci. U.S. **60**, 160 (1968).
Maas, W. K.: Cold Spring Harbor Symp. Quant. Biol. **26**, 183 (1961).
Magana-Plaza, I., Ruiz-Herrera, J.: J. Bacteriol. **93**, 1294 (1967).
Marshall, R., Redfield, B., Katz, E., Weissbach, H.: Arch. Biochem. Biophys. **123**, 317 (1968).
Martin, R. G.: Ann. Rev. Genetics **3**, 181 (1969).
Misawa, M., Nara, T., Udagawa, K., Abe, S., Kinoshita, S.: Agr. Biol. Chem. Tokyo **28**, 690 (1964).
— — — — — Agr. Biol. Chem. Tokyo **33**, 370 (1969).
Miyajima, R., Otsuka, S., Shiio, I.: J. Biochem. Tokyo **63**, 139 (1968).
Morse, S. A., Mah, R. A.: Bacteriol. Proc. p. 10 (1969).
Moyed, H. A.: J. Biol. Chem. **235**, 1098 (1960).
— J. Biol. Chem. **236**, 2261 (1961a).
— Cold Spring Harbor Symp. Quant. Biol. **26**, 323 (1961b).
Musilkova, M., Fencl, Z.: Folia Microbiol. Prague **9**, 374 (1964).
Nakamura, K., Gowans, C. S.: Nature **202**, 826 (1964).
Nakayama, K., Kitada, S., Kinoshita, S.: J. Gen. Appl. Microbiol. Tokyo **7**, 52 (1961).
— Suzuki, T., Sato, Z., Kinishita, S.: J. Gen. Appl. Microbiol. Tokyo **10**, 133 (1964).
— Tanaka, K., Ogino, H., Kinoshita, S.: Agr. Biol. Chem. Tokyo **30**, 611 (1966).
Neidhardt, F. C.: J. Bacteriol. **80**, 536 (1960).
Newell, P. C., Tucker, R. G.: Biochem. J. **98**, 39 p. (1966).
Novick, A., Horiuchi, T.: Cold Spring Harbor Symp. Quant. Biol. **26**, 239 (1961).
Novick, R. P.: Bacteriol. Rev. **33**, 210 (1969).
Oakberg, E. F., Luria, S. E.: Genetics **32**, 249 (1947).
O'Donovan, G. A., Gerhart, J. C.: Bacteriol. Proc. 125 (1968).
Ohta, Y., Ueda, S.: Appl. Microbiol. **16**, 1293 (1968).
Okumura, S., Shibuya, M., Konishi, S., Ishida, M., Shiro, T.: Agr. Biol. Chem. Tokyo **28**, 742 (1964).
Papiska, H. R., Lichstein, H. C.: J. Bacteriol. **95**, 1173 (1968).
Pardee, A. B.: In: Fermentation Advances. Perlman, D. (Ed.), p. 3. New York: Academic Press, Inc. 1969.
Patte, J. C., Le Bras, G., Loviny, T., Cohen, G. N.: Biochim. Biophys. Acta **67**, 16 (1963).
Perlman, R., Pastan, I.: Biochem. Biophys. Res. Commun. **37**, 151 (1969).
Pruess, D. L., Johnson, M. J.: J. Bacteriol. **94**, 1502 (1967).
Ramakrishnan, T., Adelberg, E. A.: J. Bacteriol. **87**, 566 (1964).
Reese, E. T., Lola, J. E., Parrish, F. W.: J. Bacteriol. **100**, 1151 (1969).
Reh, M., Schlegel, H. G.: Arch. Mikrobiol. **67**, 110 (1969).
Reyes, P., Chichester, C. O., Nakayama, T. O. M.: Biochim. Biophys. Acta **90**, 578 (1964).
Romano, A. H., Kornberg, H. L.: Biochim. Biophys. Acta **158**, 491 (1968).
Roth, J. R., Hartman, P. E.: Virology **27**, 297 (1965).
Rowbury, R. J.: Nature **206**, 962 (1965).
Sasaki, T.: J. Gen. Appl. Microbiol. Tokyo **11**, 203 (1965).
Scherr, G. H., Rafelson, M. E.: J. Appl. Bacteriol. **25**, 187 (1962).
Sheperdson, M., Pardee, A. B.: J. Biol. Chem. **241**, 5886 (1960).
Sheppard, D. E.: Genetics **50**, 611 (1964).
Shiio, I., Nakamori, S.: Agr. Biol. Chem. Tokyo **33**, 1152 (1969).
Sirotnak, F. M., Hatchell, S. L., Williams, W. A.: Genetics **61**, 313 (1969).
Spaern, U., Froholm, L. O., Laland, S. G.: Biochem. J. **102**, 586 (1967).

Stadtman, E. R.: Advan. Enzymol. **28**, 41 (1966).
— Ann. N.Y. Acad. Sci. **151**, 516 (1968).
Teteryatnic, A. F., Bryzgalova, L. S.: Antibiotiki **13**, 252 (1968).
Tomino, S., Yamada, M., Itoh, H., Kurahashi, K.: Biochemistry **6**, 2552 (1967).
Tomonaga, G.: J. Gen. Appl. Microbiol. Tokyo **12**, 267 (1966).
Trenina, G. A., Trutneva, E. M.: Antibiotiki **11**, 770 (1966).
Volkova, L. P., Butenko, S. A., Kenig, E. G.: Prikl. Biokhim. i Mikrobiol. **1**, 420 (1965).
Walker, J. B., Hnilica, V. S.: Biochim. Biophys. Acta **89**, 473 (1964).
Weinberg, E. D.: In: Antibiotics. Vol. II, Biosynthesis. Gottlieb, D. and Shaw, P. D. (Eds.), p. 240. New York: Springer-Verlag 1967.
— Advan. Microbiol. Physiol. **4**, 1 (1970).
— Tonnis, S. M.: Can. J. Microbiol. **13**, 614 (1967).
Wilson, A. C., Pardee, A. B.: J. Gen. Microbiol. **28**, 283 (1962).
Woodruff, H. B.: Symp. Soc. Gen. Microbiol. **16**, 22 (1966).
Yarmolinsky, M.: In: Viruses, Nucleic Acids, and Cancer, p. 151. Baltimore: Williams and Wilkins Co. 1963.
Zickler, F.: Abstracts, 3rd International Fermentation Symposium, New Brunswick, N.J. (1968).

A. L. Demain
Fermentation Microbiology Laboratory
Department of Nutrition and Food Science
Massachusetts Institute of Technology
Cambridge, Massachusetts 02139 U.S.A.

CHAPTER 6

Novel Energy and Carbon Sources

A. The Production of Biomass
from Hydrogen and Carbon Dioxide

H. G. Schlegel and R. M. Lafferty

With 12 Figures

Contents

1. Introduction

Food production is still based on agriculture. The carbon skeletons of all nutritional compounds are synthesized from the carbon dioxide of the atmosphere by the photosynthesis of green plants. Fossil carbon, the product of prehistoric photosynthesis, has hitherto been used only as an energy source and for the production of organochemical compounds.

Since the discovery by Just *et al.* (1948, 1951, 1952) that yeasts are able to grow reasonably fast on hydrocarbons, there has been an increasing interest in using petroleum or "liquefied" coal as carbon and energy sources for food production (Humphrey, 1967).

Hydrogen bacteria, which like green plants use carbon dioxide as a carbon source, and which use molecular hydrogen as source of reducing power and energy, have not as yet been seriously considered as potential sources of food on this planet (Mateles and Tannenbaum, 1968; Tannenbaum *et al.*, 1966), although since 1963 (Chapman *et al.*, 1963; Schlegel, 1964) their application in a bioregenerative system for space ships and on the moon has been discussed. Enormous amounts of carbon dioxide are produced during the combustion of fossil carbon and molecular hydrogen can readily be produced from water by electrolysis. With the exploitation of new energy sources – hydrodynamic processes, nuclear fission and possibly fusion reactions – the hydrogen bacteria gain increasing interest as potential sources of biological material having food value, especially protein. This chapter is based on a report given in 1968 (Schlegel, 1969) and takes into consideration some experimental results obtained more recently.

2. The Aerobic Hydrogen Bacteria

There are several groups of hydrogen oxidizing bacteria which theoretically lend themselves to the production of biomass. Many phototrophic bacteria are able to use molecular hydrogen as a hydrogen donor and carbon dioxide as a carbon source. However, all strains so far investigated are obligate phototrophs (Klemme and Schlegel, 1969). Although some strains are able to grow aerobically in the dark with an organic substrate as well as anaerobically in the light utilizing hydrogen and carbon dioxide, they do not grow aerobically in the dark with H_2 and CO_2. Neither *Clostridium aceticum*, which can be grown in enrichment culture with hydrogen and carbon dioxide, nor methanogenic bacteria are potential candidates for biomass production since their growth rates are rather modest.

The aerobic "hydrogen bacteria", generally known as "Knallgasbacteria", are not phototrophic and are very easy to grow. The formation of cell material from hydrogen, oxygen, and carbon dioxide can be summarized by the following equation:

$$6\,H_2 + 2\,O_2 + CO_2 \rightarrow \langle CH_2O \rangle + 5\,H_2O.$$

Most hydrogen bacteria are rodlike and are known under the generic name *Hydrogenomonas* (Fig. 1). Several strains were given specific names (*Hydrogenomonas eutropha, H.facilis, H. ruhlandii*), others were designated only by a strain number (*Hydrogenomonas H 16, H 1, H 20*). They may belong to several different taxonomical entities. *Micrococcus denitrificans*, some mycobacteria and actinomycetes have also been described as being able to grow like hydrogen bacteria. Several strains produce carotenoid pigments and most of the rodlike bacteria accumulate large amounts of poly-*β*-hydroxybutyric acid as an intracellular

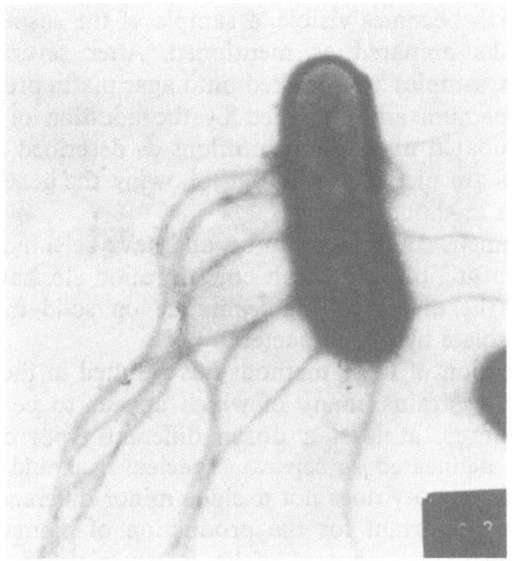

Fig. 1. *Hydrogenomonas* strain *H 16* (Wilde, 1961) which is similar to *Hydrogenomonas eutropha*. Electron micrograph

reserve material. All the strains mentioned are mesophilic bacteria with an optimum growth temperature within the range 30 to 35° C. A thermophile has been isolated, *Hydrogenomonas thermophilus*, growing optimally at 50° C (McGee *et al.*, 1967).

3. Enrichment and Growth

Many strains of hydrogen bacteria have been isolated from simple enrichment cultures. A nutrient solution, which can also be used for the growth of mass cultures, contains the following ingredients per liter: 1 g NH_4Cl; 9.0 g $Na_2HPO_4 \cdot 12 H_2O$; 1.5 g KH_2PO_4; 0.2 g $MgSO_4 \cdot 7 H_2O$; 5 mg $FeNH_4$-citrate; 10 mg $CaCl_2 \cdot 2 H_2O$; 0.5 g $NaHCO_3$;

1000 ml distilled water. Normally, 100 ml Erlenmeyer flasks each containing 15 ml of nutrient medium are inoculated with a gram of soil or with a water sample and then placed in an anaerobic jar or simply in a desiccator. After evacuation, the container is filled with a gas mixture of 70% hydrogen, 20% oxygen and 10% carbon dioxide. The container is normally incubated at 30° C and may be shaken on a shaking machine. After a few days, the medium becomes turbid or a pellicle is formed. When bacteria which form homogenous suspensions are required, the use of shaken cultures is indicated (Wilde, 1961). When bacterial growth becomes visible, a sample of the suspension is transferred to flasks prepared as mentioned. After several transfers to liquid cultures, samples are streaked onto agar plates prepared from the same liquid medium and solidified by the addition of 2% agar. The plates are incubated under the conditions as described above. Finally, single colonies are picked and tested following the general procedures of pure culture methods.

The enrichment conditions can be, and have been modified in many ways. Temperature, pH-value, salt concentration etc. have been varied. Furthermore, the direct plating technique on solid media has been employed to isolate hydrogen bacteria.

The application of these methods has resulted in the isolation of a large number of strains, many of which appear to be taxonomically identical. However, at least a dozen different types do exist which deserve to be delineated as separate species. It should be mentioned that taxonomic identity does not exclude minor differences in qualities which may be important for the production of biomass of nutritive value, e.g. growth rate, content of lipids, nucleic acids and lipopolysaccharides, protein quality etc.

4. Growth Requirements

The growth requirements of hydrogen bacteria appear to be simple. As far as has been investigated, they utilize ammonium, nitrate or urea as nitrogen sources, and require potassium, magnesium, calcium, phosphate, sulfate, iron and probably nickel in the mineral nutrient medium. The pH has to be maintained at 6.5–7.5 and is kept constant by the CO_2-bicarbonate buffer system. *Hydrogenomonas H 16*, the strain usually investigated in our laboratory, and some other strains as well, do not require chloride. For this reason, *Hydrogenomonas* can be grown in nutrient media in which Knallgas is produced directly by the electrolysis of the nutrient solution employing platinum electrodes (Schlegel and Lafferty, 1964, 1965).

The most critical conditions which influence growth do not pertain to the nutrient solution but to the gaseous components. The partial pressure of oxygen in the gas atmosphere cannot be varied ad libitum. Usually, a gas mixture containing 20% oxygen is used for growth in submerged culture as well as on solid media. The oxygen content of this mixture seems to be optimal for densely grown suspensions.

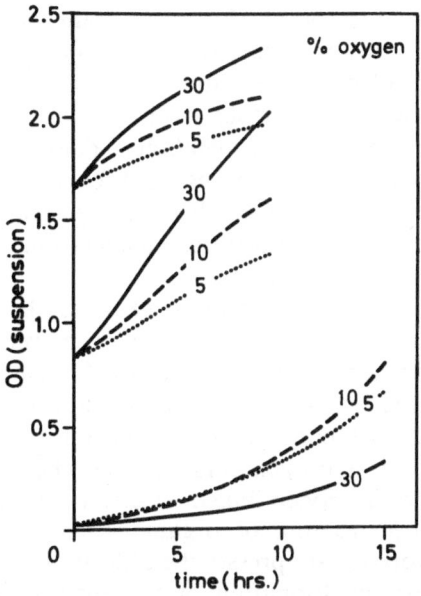

Fig. 2. Growth of *Hydrogenomonas H 16* at different oxygen concentrations starting with different cell densities. 3 l mineral medium pH 6.8 in 6 l flasks; magnetic stirring 650 rpm; gas atmosphere 5, 10 or 30% oxygen + 10% carbon dioxide + molecular hydrogen ad 100%. Turbidity measurements at 436 nm (from Schlegel *et al.*, 1961)

However, the growth of the single cell is impaired by such high oxygen partial pressures; it grows best at 0.03 atmospheres of oxygen (Fig. 2). If agar plates inoculated with ca. 10^8 cells are incubated under an oxygen hydrogen mixture containing 20% oxygen, each cell will form a colony; with 60% oxygen there is no growth. Among 40 strains thus far examined, only two tolerated such a high oxygen partial pressure. However, if in a submerged culture the density of the cell population is rapidly increasing and growth becomes limited by the rate of oxygen diffusion from the gas phase into the liquid phase, a higher partial pressure of oxygen will improve growth.

10*

5. Basic Metabolism of Hydrogen Bacteria

Hydrogen bacteria are facultative chemoautotrophs; they grow as well or at an even higher rate under heterotrophic conditions. Fructose is the only carbohydrate which many strains such as *Hydrogenomonas eutropha* and strain *H 16* can utilize as a substrate. Most strains are able to grow with organic acids and amino acids. Some strains are also able to decompose aromatic compounds, purines, and pyrimidines.

The uptake of fructose occurs by active transport, while glucose, which can be utilized only by rare mutants, enters the cell by diffusion only. Fructose is degraded via the Entner-Doudoroff pathway and the products of this pathway are further oxidized via the usual routes of intermediary metabolism. The basic metabolism does not differ from that of other aerobic bacteria. The strains mentioned are even able to respire anaerobically with nitrate as a hydrogen acceptor; they grow as denitrifiers with fructose as the substrate. No anaerobic growth has been observed under autotrophic conditions.

Autotrophic carbon dioxide fixation is catalyzed by the enzymes of the reductive pentose phosphate cycle, reducing power being supplied by NAD-linked hydrogenase and ATP being furnished by oxidative phosphorylation. Some enzymes involved in autotrophic growth are inducible, others are fairly constitutive. Under autotrophic conditions, i.e. in the presence of hydrogen and oxygen, the NAD-dependent hydrogenase and the hydrogen-linked electron transport chain are formed even when e.g. fructose or organic acids are present in the nutrient medium. Hydrogen suppresses the utilization of a great number of organic substrates; the formation of the enzymes of the Entner-Doudoroff pathway is repressed in the presence of hydrogen as H donor. In addition, the utilization of fructose is inhibited by hydrogen; this inhibition is caused by the ATP and eventually the NADH produced during hydrogen oxidation; ATP acts as an allosteric inhibitor of glucose-6-phosphate dehydrogenase which is involved in fructose degradation (Blackkolb and Schlegel, 1968a, b). These control mechanisms, which have been studied in *Hydrogenomonas H 16*, result in the dominance of autotrophic metabolism as long as hydrogen and carbon dioxide are available and exert a "sparing" effect on the consumption of organic substrates.

6. Growth in Submerged Culture

For the growth of hydrogen bacteria under autotrophic conditions in submerged mass culture, the nutrient solution described above may be used; only ammonia may become growth-limiting and an additional amount has to be supplied when a cell density of 8 g dry weight/liter has been reached. Furthermore, the pH has to be controlled since some strains do eventually excrete organic acids.

a) Batch Culture Systems

Methods to cultivate knallgasbacteria in mass culture have been developed slowly. They are the result of a compromise; they should make possible high growth rates, but they should also guarantee sufficient safety. The danger is already suggested by the name knallgas-bacteria, knallgas being the explosive mixture of hydrogen and oxygen burning with a "Knall" (= "bang"). In order to avoid any mechanical friction caused by moving parts and discharges of static electricity in the gas space of the vessels, we first used a magnetic stirring device. Flat-bottomed round flasks with a volume of 6 liters were half filled with nutrient solution and connected to a container of the appropriate gas mixture. When magnetically stirred at 600 r.p.m., the gas diffusion was sufficient to support cell growth up to a density of 2–3 g cell dry weight/l. For these experiments the gaseous components were premixed in the gas container and the mixture was then sucked over into the culture

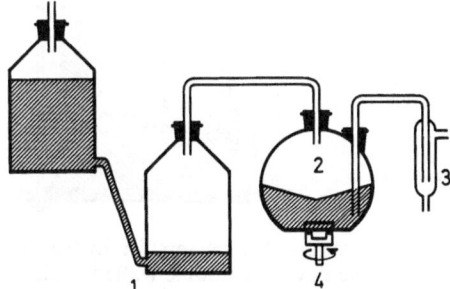

Fig. 3. Gascontainer system for cultivating *Hydrogenomonas* under a gas atmosphere of hydrogen, oxygen and carbon dioxide. *1* gas reservoir; *2* culture vessel; *3* sampling device; *4* magnetic stirrer

vessel as it was consumed (Fig. 3). In these vessels the cells did not grow exponentially beyond 1 g dry weight/l. In order to obtain higher gas diffusion rates and consequently higher cell densities, two other principles for gassing the cultures were introduced: a closed circuit system and a constant flow device. In both cases the culture vessels already described were used.

In the closed circuit system, a stream of gas is pumped from a gas container into the culture vessel and then back to the container

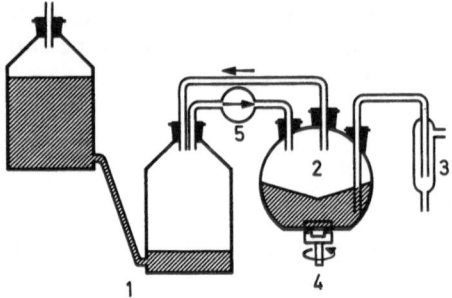

Fig. 4. Closed circuit system for cultivating *Hydrogenomonas*. *1* gas reservoir; *2* culture vessel; *3* sampling device; *4* magnetic stirrer; *5* pump

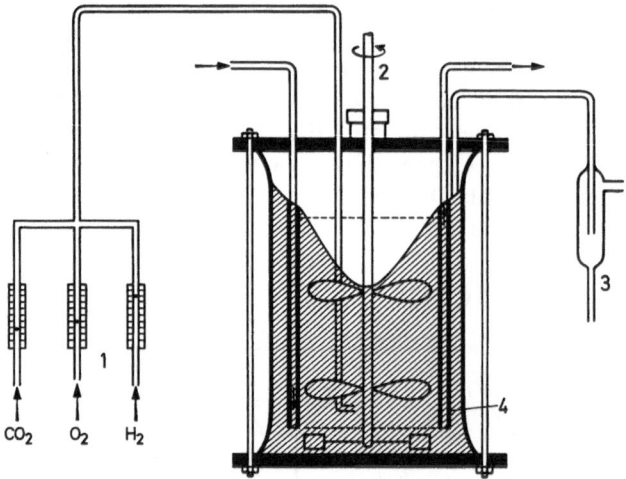

Fig. 5. Continuous flow system. *1* flow meters for hydrogen, oxygen and carbon dioxide; *2* stirrer; *3* sampling device; *4* double walled Waldhof cylinder connected to a thermostat

(Fig. 4). With this system the exponential growth rates were maintained for a longer period, resulting in higher densities and greater yields (4 g dry weight/l).

In the continuous flow system (Fig. 5) the gases were taken from steel tanks and led via reducing valves, flow meters and a mixing chamber to the culture vessel. The highest yields were obtained with this system. However, the exponential growth ceased before the culture had reached 10 g dry weight/l.

As already mentioned, suspensions of high densities consume large volumes of gas. The normal gas consumption rate of *Hydrogenomonas*

H 16 has been measured in Warburg vessels (Eberhardt, 1965). It amounts to ca. 3000 µl H_2-O_2-CO_2-mixture/hour/mg protein, when the partial pressure of oxygen is equal to that in the air (0.2 at). This gas uptake rate corresponds approximately to an oxygen demand rate of 0.66 l O_2/hour/g dry weight. A suspension of 10 g cell dry weight per liter will, therefore, consume 6.6 l oxygen or nearly 300 mmoles oxygen per hour.

This value corresponds to oxygen absorption rates which have been measured in shaken flasks and fermenters under optimum conditions. In our fermenters, a maximum oxygen absorption rate of 250 mmoles O_2/hour/l was achieved when a strong stream of air was supplied and the solution was stirred at 2000 r.p.m. The determination of the oxygen transfer rates was made by the sulfite oxidation method.

From a comparison of these data, it can be concluded that the growth of a culture growing in a fermenter, which has an oxygen absorption rate of 200 mmoles O_2/hour/l, is restricted by the oxygen supply. In this context, older experimental results are of interest: the growth of *Hydrogenomonas* is still exponential, even when the oxygen supply rate satisfies only half of the oxygen demand (consumption) rate and then declines (Schlegel and Lafferty, 1965). This means that the maximum hydrogen oxidation rate exhibited by the cells in Warburg vessels represents a luxury oxidation rate, and the energy requirement for normal growth is satisfied with half this value.

These considerations explain why, in a fermenter with a normal impeller and aeration system, exponential growth ceases at 10 to 15 g cell dry weight. Further growth continues, but at an ever-decreasing rate.

b) Continuous Culture Systems

A continuous culture apparatus for *Hydrogenomonas* with an external gas supply has been described (Foster and Litchfield, 1964). The authors used a culture vessel designed to accommodate about 2 l of cell suspension with its associated gas phase. The vessel consisted of a glass cylinder closed at each end with stainless steel plates. A violent agitation was achieved by peripheral baffles and by a high speed stirrer which could be driven up to 1500 r.p.m. A reservoir vessel containing the sterile nutrient solution and a storage vessel for the harvested cell suspension were connected to the culture vessel using the usual connection tubes.

The apparatus is characterized by a highly developed control system. The gas supply is controlled by three immersed sensors, specifically sensitive to each of the gases. When the sensor signal falls below a

preset concentration, a definite increment of gas is added periodically
until the dissolved concentration reaches the proper value. The cell
density is maintained at a constant level by dilution with the nutrient
medium. The pH-value and temperature are also kept constant. With
respect to the growth rate, cell density, and yield, no quantitative data
have been reported so far. Since the gas supply control system is based
on the maintenance of preset measurable gas concentrations in the
nutrient medium, the cell density must have been rather low.

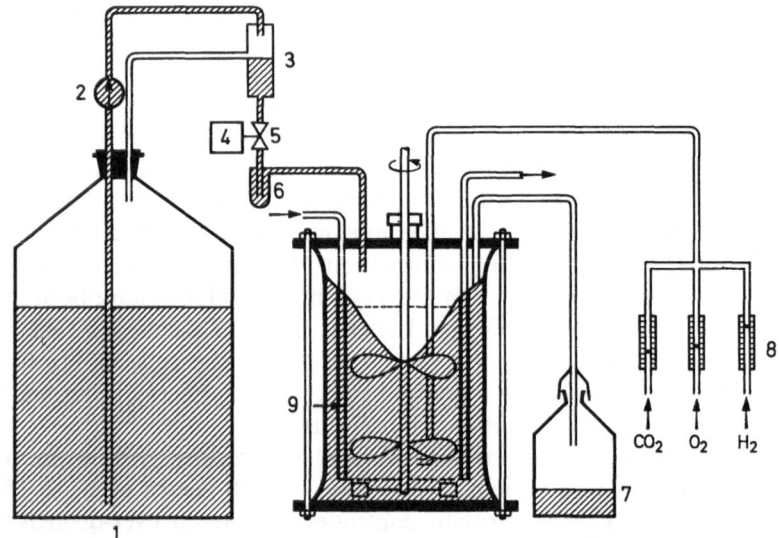

Fig. 6. Chemostat for continuous cultivation of *Hydrogenomonas*. *1* medium
reservoir; *2* pump; *3* constant-level vessel; *4* timer; *5* valve; *6* thermal trap;
7 collecting vessel; *8* flow meters; *9* Waldhof cylinder

A similar continuous culture system accommodating 4 l liquid
medium is in use in our laboratory for permanent production of cells
needed for different kinds of experiments (Fig. 6). The vessel contains a
double-walled Waldhof cylinder, which serves for heat transfer, and a
stirrer which can be driven up to 2000 r.p.m. The flow of the incoming
medium is controlled by a magnetic valve with timer and passes a
thermal-trap heated to 70° C in order to prevent backgrowth of cells
into the reservoir vessel. The gas supply is controlled by flow meters.
With this system, a constant cell density of 3 g dry weight/l could be
maintained for months.

An experimental continuous culture system with an internal gas
supply has been developed by Schuster (Schlegel *et al.*, 1967; Schuster

and Schlegel, 1967). The kinetics of growth have been studied only at low
cell densities. By means of an electrode-triplet consisting of one cathode
and two anodes (one external to the culture), the production rates of
hydrogen and oxygen within the suspension could be varied independently of each other within a wide range. Growth parameters,
enzyme activities and other properties of the cells have been studied in
continuous culture with limiting concentrations of hydrogen and
oxygen (Table 1).

Table 1. *Comparison of the growth of Hydrogenomonas H 16 in the chemostat with
limiting concentrations of hydrogen or oxygen (from Schuster and Schlegel, 1967)*

Parameters		Limitation	
		by H_2	by O_2
Culture volume V	(ml)	640	725
Growth rate	(h^{-1})	0.081	0.080
Cell density	(mg/ml)	0.167	0.237
PHB content	(% dry weight)	< 1	23
Ratio H_2/CO_2		9.1	4.6

When the growth rate is limited by oxygen, the cells accumulate
PHB. With limiting concentrations of hydrogen, the PHB content is
negligible. As can be concluded from the H_2/CO_2-ratio, the growth
yield under oxygen limitation is twice as high as under hydrogen
limitation. From these measurements it can be concluded that in order
to obtain a highly efficient conversion of hydrogen to biomass, the cells
should be grown with oxygen as the growth limiting factor.

c) The Admixture of Two Gases: a Novel Problem

The prerequisites for an anaerobic fermentation are relatively easy to
fulfil since all the components necessary for anaerobic microorganisms
can be added to the nutrient medium before inoculation. For the
growth and fermentation of aerobic microorganisms one component –
oxygen – has to be administered continuously because of its low
solubility in the aqueous nutrient medium. Therefore, aerobic fermentations in submerged culture are much more difficult to carry out than
anaerobic fermentations. In dense suspensions, the oxygen demand rate
may even exceed the oxygen absorption rate, and growth will be
limited by the rate of gas transfer into the medium. Although much
effort has been invested, the development of aeration systems is not
yet complete and requires further improvement.

New problems arise when not only the hydrogen acceptor – oxygen – but also the energy sources are to be added in the form of gases. In this case, at least two nutrient components have to be added continually, one of which may limit growth. Two gases, for example, have to be added in order to grow bacteria which oxidize methane or other gaseous hydrocarbons; hydrogen bacteria even require three gases. The question is, in what ratio should the gases be mixed. Up to the present, gas mixtures were used containing the component gases in proportions which are stoichiometrically related to the ratio in which the gases are consumed. This procedure did not take into consideration that the solubility and the absorption rates of one gas may differ from those of the other gas components.

Therefore, the relative amounts of the gases supplied to the culture should be adjusted in such a way that there is a direct proportionality between the absorption rates and the consumption rates. This means, that e.g. if hydrogen has a low absorption rate, a high hydrogen partial pressure should be employed in order to balance the absorption rates of hydrogen and oxygen.

Unfortunately, neither the relative ratio of the absorption rates of hydrogen and oxygen is not known nor are practicable methods available to measure the absorption rate of hydrogen in aqueous solutions. A method for determining hydrogen absorption rates analogous to the sulfite-oxidation method for oxygen has still to be developed.

However, even if the theoretical absorption rates of hydrogen relative to oxygen were known, it would not be permissible to draw conclusions pertaining to the actual absorption rates in a bacterial suspension. Oxygen absorption rates are influenced by the presence of proteins or detergents in the medium (Tsao, 1968); presumably the air bubbles are surrounded by mono-molecular protein layers. Other components of the nutrient medium may influence the gas absorption rates (cellular debris, cell wall lipopolysaccharides, slimes etc.). It is highly uncertain whether the diffusion rates of hydrogen and oxygen are equally influenced by these factors. Therefore, we prefer direct determinations of oxygen (and hydrogen) in suspensions of hydrogen bacteria during "aeration" with different H_2/O_2-mixtures.

When the oxygen partial pressure is kept constant (0.1 resp. 0.2 at) and that of hydrogen is varied, the amount of dissolved oxygen which can be measured using an oxygen electrode will change. With a suspension of medium cell density (1 g dry weight/l) and a stream of air (0.2 at O_2), the dissolved oxygen measured is 7.4 mg O_2 per liter at 30° C. The decrease in the oxygen content compared to a bacteria-free medium (7.6 mg O_2) is due to the endogenous respiration of the cells. When hydrogen is added to the gas stream and the oxygen partial pressure

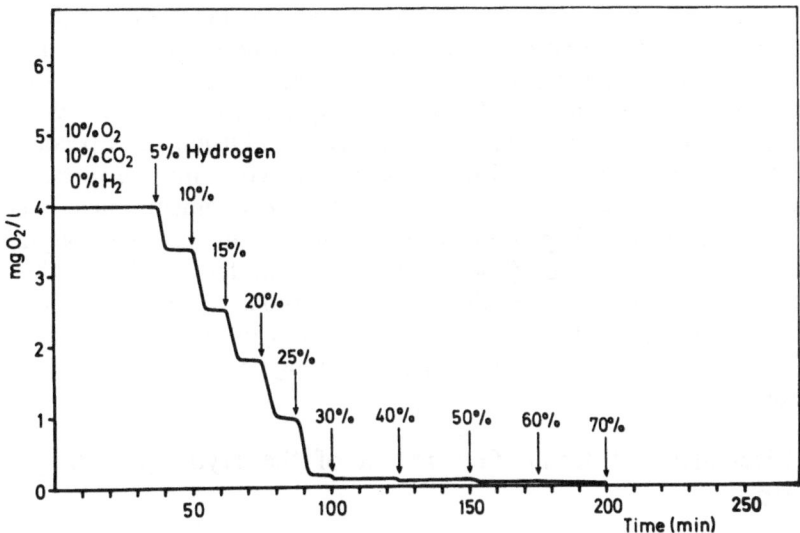

Fig. 7. Reduction of dissolved oxygen concentration in a suspension of *Hydrogenomonas H 16* stirred under an atmosphere of 10 per cent oxygen, 10 per cent carbon dioxide, and varying concentrations of hydrogen. 4.5 g cells dry weight per liter; gas flow rate 300 ml/min; temperature 30° C; 700 rpm. Oxygen was measured using a polarographic electrode

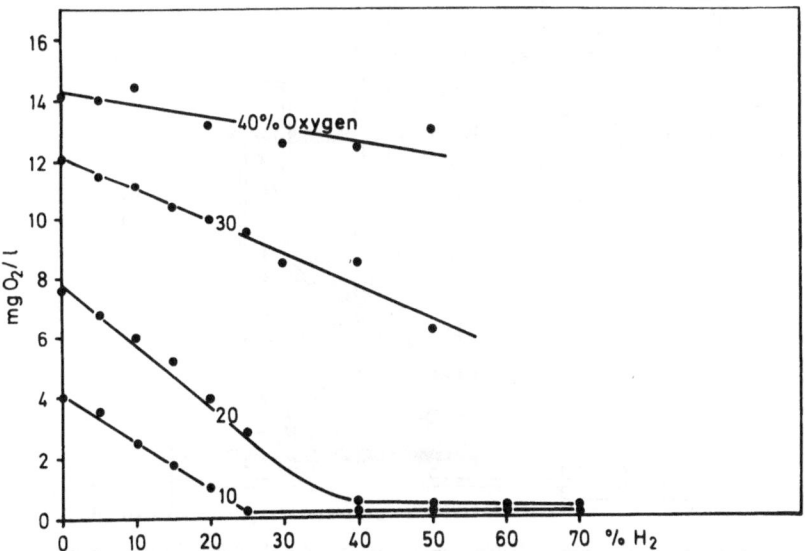

Fig. 8. The influence of the hydrogen concentration on the dissolved oxygen concentration measured in a suspension of *Hydrogenomonas H 16* stirred under the conditions, outlined in Fig. 7, at four different oxygen concentrations in the gas phase

held constant, the consumption rate of oxygen increases drastically resulting in a pronounced decrease of the concentration of dissolved oxygen. With increasing partial pressures of hydrogen, still more oxygen is consumed and the amount of dissolved oxygen decreases further. When the relative ratio of the hydrogen and oxygen diffusion rates becomes equal to the ratio of the hydrogen and oxygen consumption rates, the dissolved oxygen value becomes zero. These experiments have been performed using cell suspensions with 4.5 g dry weight/l and with varying oxygen partial pressures (Fig. 7 and 8). From the results of these experiments it can be concluded that the gaseous components should be mixed in a ratio of approximately 2 : 1 for hydrogen and oxygen in order to satisfy the gas consumption of a growing population of hydrogen bacteria.

d) Electrolysis: Internal Generation of the Hydrogen Oxygen Mixture

With the culture methods thus far described, hydrogen and oxygen are supplied to the culture from external sources such as gas tanks or an electrolysis apparatus. It was found in 1963 that the oxygen-hydrogen

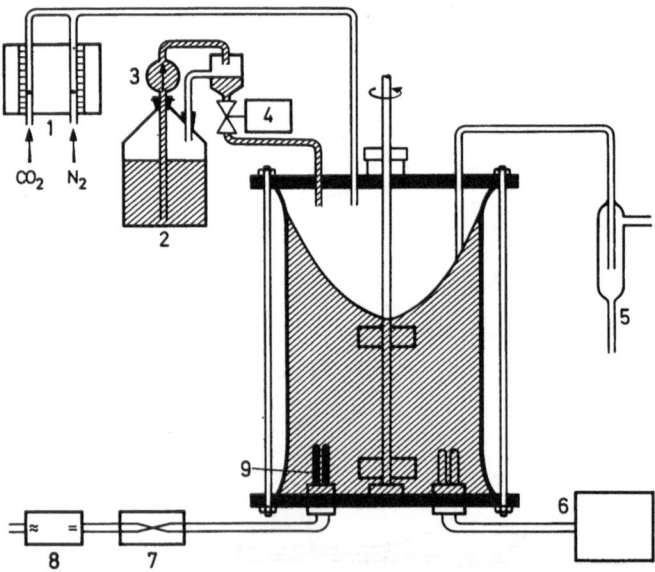

Fig. 9. Apparatus for growing *Hydrogenomonas H 16* in batch culture with internal gas supply (electrolysis of the culture medium by platinum electrodes). Carbon dioxide containing nitrogen is blown through the air space of the vessel. Auxiliaries are shown. *1* flow meters; *2* solution of iron salts; *3* pump; *4* timer; *5* sampling device; *6* temperature control; *7* polarity switch; *8* rectifier; *9* electrodes

mixture can be produced by the direct electrolysis of the mineral nutrient solution in the culture vessel using platinum electrodes (Schlegel and Lafferty, 1964, 1965). Since chlorides can be omitted from the medium, no chlorine is developed and it is possible to grow *Hydrogenomonas H 16* using direct current to supply energy; only carbon dioxide is taken from an external source. The electrolytic production of "Knallgas" by means of platinum electrodes is undoubtedly the simplest method for supplying the bacteria with both hydrogen and oxygen (Fig. 9). Since hydrogen and oxygen are both produced and consumed within the nutrient medium, there is no danger of explosion. A fermenter with internal generation of "Knallgas" has been designed and built by Rudolph (1968). For unknown reasons, the cell density in this vessel employing electrolytically produced hydrogen and oxygen never exceeded 2.8 g dry weight/l.

7. Nutritive Value of Bacteria, Especially Hydrogen Bacteria

The gross composition of the basic cell substances of all bacteria is very similar. On an average, bacteria consist of 50% protein, 15% nucleic acids and 20% cell wall substances. However, the composition differs with respect to storage materials and cell wall and capsular substances.

The biological value of bacterial proteins for animal nutrition appears to be good, as far as has been demonstrated using *Escherichia coli* as an experimental diet for mice and rats (Roberts, 1953). In addition, the amino acid composition of *Azotobacter chroococcum, A. microcytogenes* and bacilli is indicative of a high nutritive value (Nilsson and Jönsson, 1962; Mateles *et al.*, 1967). The composition of the bacterial protein, as given in Table 2, and its nutritive value resemble the data for casein (Ogur, 1966). This conclusion has been confirmed by feeding cells of *Hydrogenomonas eutropha* to rats (Waslien and Calloway, 1967; Calloway and Kumar, 1969). The concentrations of the amino acids lysine, methionine, threonine, and tryptophan were similar to those found in casein and exceeded the concentrations found in proteins of lower biological value. The biological value of casein and of the bacteria was 77%.

The composition of bacterial cells differs with respect to storage materials and capsular substances. The ability to synthesize and to accumulate intracellular storage material and extracellular slimy polysaccharides or polypeptides are intrinsic properties. The extent to which an accumulation of one or more of these substances occurs can be modified by cultural conditions. The capability to produce these materials can even be modified or eliminated by mutation.

Table 2. *Amino acid composition of Hydrogenomonas eutropha, Peudomonas saccharophila and casein (taken from Calloway and Kumar, 1969); values expressed as g/16 g of nitrogen*

| Amino acid | H. eutropha | Published compositions | | |
		Casein	H. eutropha	P. saccharophila
Tryptophan		1.34	1.05	
Threonine	4.52	4.30	2.90	5.37
Lysine	8.61	8.06	3.57	5.73
Methionine	2.69	3.10	1.54	2.03
Cystine		0.38	0.11	0.36
Isoleucine	4.58	6.59	2.92	4.14
Leucine	8.52	10.11	5.44	8.35
Phenylalanine	3.96	5.42	2.96	3.56
Tyrosine	3.26	5.86	2.41	2.32
Valine	7.13	7.44	4.08	7.55
Histidine	2.48	3.04	1.28	1.81
Arginine	8.00	4.10	4.59	5.01
Alanine	8.80	3.38	6.02	13.57
Aspartic acid	9.57	7.44	5.82	9.72
Glutamic acid	11.17	2.32	10.33	10.52
Glycine	5.47	2.00	3.72	9.65
Proline	3.46	11.82	2.77	5.59
Serine	3.47	6.69	2.42	4.64
Total	95.69	93.39	63.93	99.92

One obstacle to the utilization of bacteria as food seems to be the high nucleic acid content. However, there are as yet no conclusive data which can be regarded as having proved that a diet high in nucleic acids is harmful to animals or to humans. As is well-known, ruminants are nourished by the microorganisms which grow in the rumen. They excrete three to six times as much purine as monogastrial animals. Further investigations are necessary to study the effect of nucleic acids on humans and to decide whether or not uric acid, the endproduct of purine catabolism, is really a burden for animal metabolism.

Although the DNA/protein ratio seems to be fairly constant in bacteria, the RNA content increases with the growth rate (Herbert, 1961; Dicks and Tempest, 1966) (Fig. 10). Therefore, it will be possible to decrease the RNA content by growing the cells at a low growth rate after they have been grown exponentially at the maximum rate.

The storage materials generally encountered in bacteria are either lipids or polysaccharides. Among the pseudomonads and the photo-trophic bacteria, the storage lipid usually consists of poly-β-hydroxy-butyric acid; among the mycobacteria and actinomycetes, waxes and triglycerides seem to prevail. Polysaccharides are widely distributed among bacteria and occur in the form of starch or glycogen. Both lipids and polysaccharides are accumulated in amounts up to 70% of the bacterial dry weight. In addition, many bacteria excrete large amounts of polysaccharides into the medium.

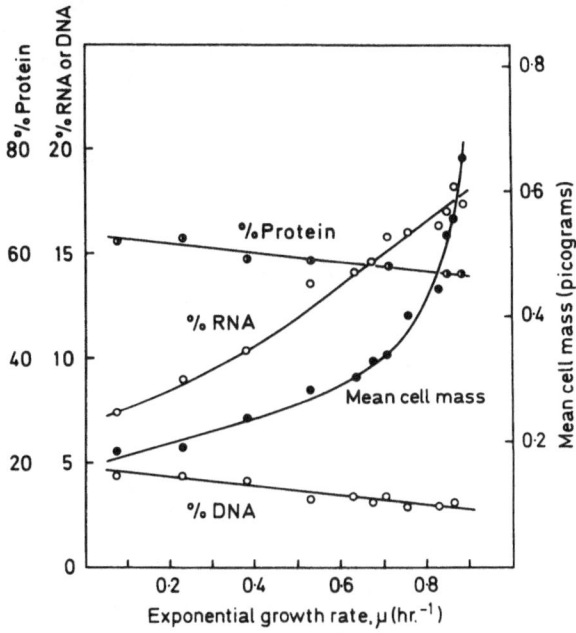

Fig. 10. Protein and nucleic acid contents of *Aerobacter aerogenes* as a function of growth rate (from Herbert, 1961)

Those hydrogen bacteria, which have been investigated so far, seem to represent several storage types. Only a few strains were found producing triglycerides, and others polysaccharides. Most strains accumulate poly-β-hydroxybutyric acid (PHB) (Fig. 11). This accumulation occurs preferentially when energy and carbon are available in excess and when growth is limited by the absence of utilizable compounds of nitrogen, sulfur or phosphorus. The lack of oxygen also results in the storage of PHB. Intracellular PHB-deposition can amount to 60% of the cell dry weight.

Our knowledge about the possible nutritive value of PHB for animals is rather fragmentary. Experiments in our laboratory using mice did not give clear results. When 50% of the diet consisted of PHB-rich *Azotobacter* cells, the mice died of obstipation. In rats the digestibility of the bacterial PHB was not more than 10%. Since the digestibility of

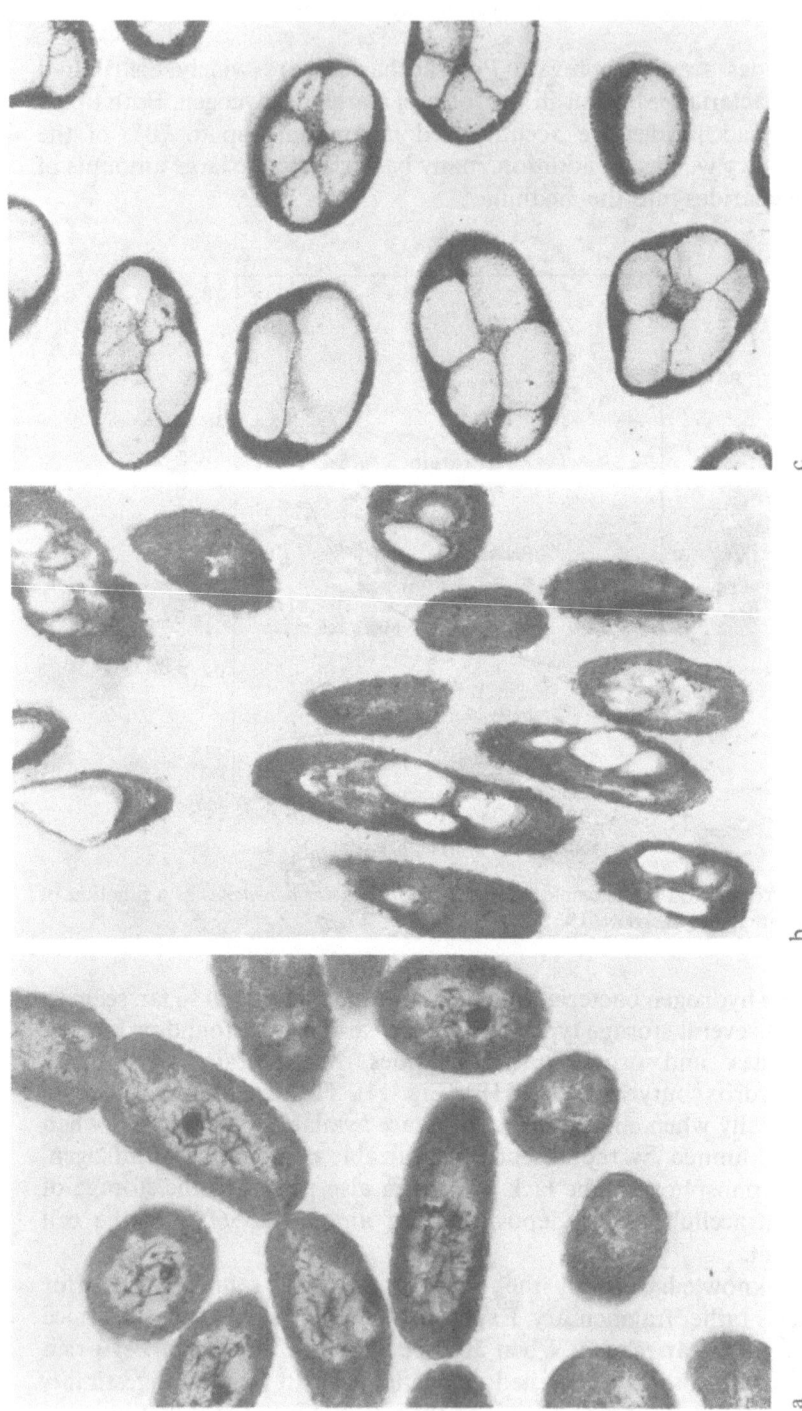

Fig. 11. Accumulation of poly-β-hydroxybutyric acid in the cells of *Hydrogenomonas H 16*. The cells had been grown exponentially (a) and were then incubated for 1 (b) and 24 hrs. (c) in the presence of 0.1% acetate as a carbon source in the absence of a nitrogen source (Electron micrographs of ultrathin sections taken by P. Hillmer and F. Amelunxen)

the bacterial protein was good (95%), bacterial cell walls are apparently not resistant to the digestive enzymes of the intestinal tract (Waslien and Calloway, 1967, 1969). Almost all the PHB is exctreted undigested. It is possible that enzymes which could depolymerize PHB are lacking in the gut and are neither produced nor excreted by intestinal bacteria. It should be mentioned that the monomeric β-hydroxybutyric acid, which can be obtained from PHB by enzymatic or alkaline hydrolysis, is an important intermediate in cellular metabolism. It is a product of the degradation of some amino acids (isoleucine), is quickly utilized in animal cells, and seems to be a better fuel for the heart muscle than glucose (Krebs, 1961). With regard to the possible use of bacteria for nutritional purposes, the digestibility of PHB, as well as the absorption and alimentary value of β-hydroxybutyric acid, should be studied more intensively.

Hydrogenomonas eutropha has already been administered to humans (Waslien *et al.*, 1969). The result of the experiments carried out with a group of six adult male volunteers is summarized by the title of the paper "Human intolerance to bacteria as food". Abdominal discomfort, headache, and weakness were the symptoms following the ingestion of 18 g washed, boiled, and lyophilized bacterial cells. However, these observations have little significance, since the bacterial preparations contained in addition to *Hydrogenomonas*, gram-positive rods, and two *Streptococcus* species as contaminants. To generalize from this experience would be like saying: "all fruits are poisonous" if deadly night shade (*Atropa belladonna*) had been the first fruit eaten!

Since bacteria make up 60 to 90% of the intestinal contents, they cannot in principle be toxic. Pathogenicity and toxicity may be caused by certain cell substances, e.g. lipopolysaccharides which vary from strain to strain. Therefore, conclusions concerning the suitability of bacteria for human nutrition can only be drawn from an extensive program involving many bacterial strains which have been grown as monocultures without contamination. Furthermore, other cell fractions, such as proteins, cell walls, cytoplasmic membranes, pure storage materials, and nucleic acids must also be tested.

8. The Isolation of New Strains and Mutants without PHB

The last paragraph can be summarized as follows: bacterial protein is comparable to milk casein (Calloway and Kumar, 1969). The nucleic acid content of bacteria depends on the growth rate and can be reduced by submitting the cells to conditions of the stationary growth phase or, when growing them in a chemostat, by growing them in a second stage. Most hydrogen bacteria synthesize and accumulate poly-β-hydroxybutyric acid which apparently has no nutritive value. Therefore, attempts were made to isolate new strains and PHB-deficient mutants.

Several enrichment conditions and isolation procedures, differing from those outlined previously, have been employed. The usual enrichment culture almost invariably results in the isolation of mesophilic pseudomonads and similar rod-like bacteria and some mycobacteria-like strains. In order to find vitamin-requiring strains, which may be present in soil or water, 90 strains were isolated from enrichment media containing 6 vitamins. However, all these strains proved to be auxoautotrophic rods (Schuster, 1967).

Since several strains of hydrogen bacteria isolated 50 years ago (Ruhland, 1922, 1924; Grohmann, 1924) were described as spore formers, and since spore formers like *Bacillus megaterium* accumulate glycogen, we tried to isolate spore-forming hydrogen bacteria from pasteurized soil (Wilde, 1961, 1962). Unfortunately, even the attempts to isolate Knallgas bacteria-forming thermoresistant spores from soil and mud samples taken from the Botanical Garden of the University of Tübingen, and from places where Ruhland might have collected his material for inoculation, failed (Eberhardt, 1965).

Carbon monoxide-oxidizing bacteria have been described (Kistner, 1953, 1954), which are also able to grow with molecular hydrogen as a substrate; therefore, enrichment cultures were prepared employing carbon monoxide as an energy source and using 72 soil and mud samples as inocula (Pfitzner, unpublished results). Of these isolates, only 8 strains were capable of growing autotrophically using molecular hydrogen as the energy source; their growth rate was rather modest. Furthermore, a propane-oxidizing mycobacterium has been found which can grow using hydrogen (Klausmeier *et al.*, 1958). On the basis of this observation, propane-enrichments were prepared. Of 100 strains isolated, only 25 strains were able to grow with hydrogen (Siebert, unpublished results). Finally, attempts were made to isolate arthrobacters able to grow as hydrogen bacteria; however, no arthrobacters were obtained.

As already mentioned, *Hydrogenomonas eutropha* and *Hydrogenomonas* strain *H 16* belong to those hydrogen bacteria having the highest growth rates and being easy to handle. Both strains accumulate PHB. Accumulation occurs when growth is limited by secondary factors. As will be described later, the efficiency of biomass production is optimal when cell growth is limited by the availability of oxygen. Under these conditions, cells are produced which contain 25% PHB. When growth is limited by hydrogen, the PHB content is negligible. However, in this case the efficiency of biomass production is also low. In order to avoid this dilemma and to be able to grow cells without

PHB with high economic efficiency, a procedure has been developed for the isolation of PHB-deficient mutants of *Hydrogenomonas H 16*. Several mutants were isolated; some of them synthesized PHB at a decreased rate and two mutants did not produce any PHB (Table 3).

Table 3. *PHB-content of mutants and the wild-type strain of Hydrogenomonas H 16 following incubation in the presence of fructose, gluconate, acetate or carbon dioxide + hydrogen in the absence of a nitrogen source (from Schlegel et al., 1970)*

Strain or mutant, resp.	Amount of PHB (per cent of dry weight) after incubation with			
	Fructose for 40 hrs	Gluconate for 23 hrs	Acetate for 23 hrs	$CO_2 + H_2$ for 26 hrs
H 16 (wild-type)	65.3	27.7	37.2	35.8
H 16 PHB⁻ 1	11.9	7.3	—	5.6
2	—	8.6	8.2	1.5
3	—	7.3	13.9	4.3
4	—	0	0	0
5	—	—	0	0

Colonies of PHB-deficient mutants can be recognized by the following procedure (Schlegel *et al.*, 1970): colonies grown on nutrient broth agar are transferred by replica plating to a nitrogen-poor fructose agar (0.5 % fructose, 0.005 % ammonium chloride); after incubation for 5 days the colonies were subjected to a lipid staining procedure. The agar plates are flooded using 10 ml of a solution of the dye Sudan black B (0.02 % in 96 % ethanol); after 20 min the dye solution is replaced by 10 ml 96 % ethanol which is removed after 1 min. The colonies of the PHB-rich wild-type cells retain the dye and appear dark blue, while the colonies of the PHB-deficient mutants are decolorized by the differentiation process and appear light gray. In order to isolate the mutants they were picked from the master plates. Enrichment methods employing the ^{32}P-phosphate inactivation technique and sucrose density gradient centrifugation have been devised (Schlegel *et al.*, 1970).

9. Production of Intermediate Metabolites and Mutant Selection

Novel energy sources such as mineral oil, natural gas and molecular hydrogen lend themselves not only for biomass production, but many substances other than cell material may also be produced. There are two ways to find strains which overproduce and excrete either some metabolite or a degradation product. One method involves the isolation of new strains and screening for excretory products; the other method is based on the selection of mutants which are defective in regulatory mechanisms or are auxotrophic.

Those strains of hydrogen bacteria so far isolated have not – or are not known to have – been adequately investigated as regards the excretion of extracellular products. In this respect, only *H. eutropha* has been examined using radioactive carbon as a tracer (Brown *et al.*, 1964). Only 2% of the total carbon assimilated appeared in the supernatant during exponential growth; during the stationary phase of growth the amount of carbon compounds excreted was 5% of the total assimilated. Ribose, glutamic acid, alanine, and tyrosine were some of the products which could be identified.

Methods to isolate mutants defective in regulatory mechanisms are well known, although modified techniques applicable to bacteria other than the *Enterobacteriaceae* and the *Bacillaceae* were lacking. Attempts to select for and to isolate auxotrophic mutants and mutants defective in regulatory mechanisms were successful, and results of recent investigations look promising (Reh and Schlegel, 1969a, b; Schlegel and Hill, 1969).

Since *Hydrogenomonas* strain *H 16* and many other strains of hydrogen bacteria are not affected by penicillin or its derivatives, the well-known penicillin technique for the selection of auxotrophic mutants cannot be applied. However, colistine (= polymyxine B) proved to be an effective agent which kills only growing cells leaving non-growing cells unimpaired (Schlegel *et al.*, 1965).

By using colistine for the enrichment procedure, many auxotrophic mutants defective in the biosynthetic pathway of valine and isoleucine have been isolated. From an isoleucine-requiring mutant, defective in threonine desaminase, a prototrophic revertant has been isolated. The threonine desaminase of this revertant differs from the wild type enzyme in that its affinity for isoleucine is diminished . This revertant excretes isoleucine. Another revertant of an isoleucine-deficient mutant was obtained which formed the enzyme acetohydroxy acid synthetase constitutively. During heterotrophic growth with fructose or lactate as substrates, valine, isoleucine and leucine were excreted into the culture medium. Approximately 0.6 g of amino acids were produced per liter suspension when lactate was supplied as a substrate; under autotrophic conditions the excretion was negligible (Reh, 1970; Fig. 12).

Different types of mutants were obtained by employing the antimetabolite technique. The growth of strain *H 16* is inhibited by $5 \cdot 10^{-5}$ M trifluoroleucine. Among the mutants which were resistant to this antimetabolite, some excretors have been found; some of these

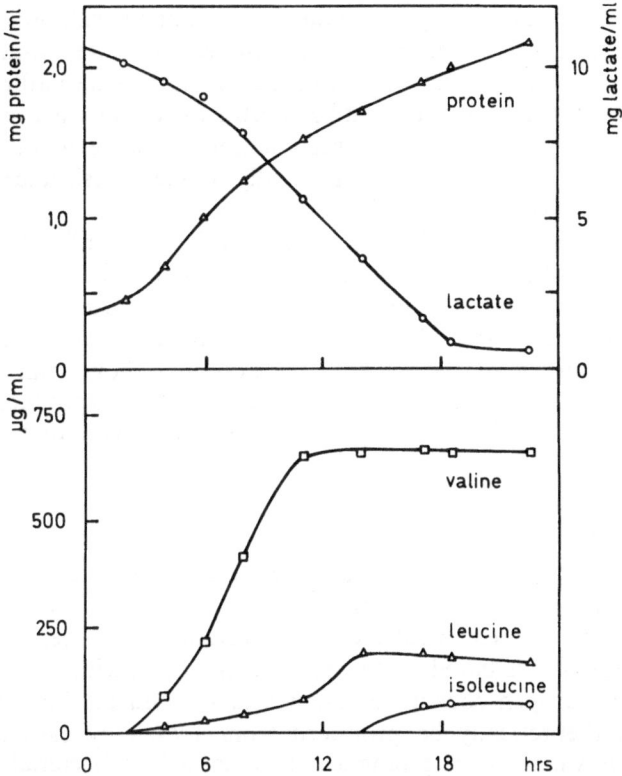

Fig. 12. Excretion of amino acids by the mutant strain *R 3* of *Hydrogenomonas H 16* during growth in a lactate-containing mineral medium. Amino acids were measured quantitatively after paper chromatography in water saturated n-butanol (from Reh, 1970)

produced satellite growth due to the excretion of leucine; others excreted both isoleucine and valine. At least four different types of mutants resistant to trifluoroleucine have been isolated and investigated (Hill and Schlegel, 1969a, b).

As expected, among the mutants excreting leucine there was a mutant constitutively derepressed with respect to the formation of the enzyme α-isopropylmalate synthetase, which is the first enzyme in the leucine biosynthetic pathway. In another mutant, this enzyme is insensitive to endproduct inhibition by leucine. However, contrary to our expectations, we found mutants carrying regulatory defects in the control of the valine-isoleucine biosynthetic pathway; several mutants are constitutively derepressed with respect to the formation of aceto-hydroxy acid synthase; and in one mutant this enzyme is insensitive to endproduct inhibition by valine. The selection and the existence of these mutant types among the mutants resistant to trifluoroleucine is due to an unusual property of the α-isopropylmalate synthetase of strain *H 16*. L-valine is not only related to the leucine pathway as a precursor, but also acts as a positive effector for the first enzyme of the leucine pathway. Hence, the overproduction of valine is followed by overproduction of leucine which finally confers resistance to the bacteriostatic effect of trifluoroleucine.

Although the yields of excreted metabolites are still rather small under autotrophic conditions, they can possibly be improved by modifying the regulatory mechanism of catabolic reactions. Mutants constitutive in glucose-6-phosphate dehydrogenase have already been obtained (König *et al.*, 1969). As these examples show, strain *H 16* is accessible to many manipulations resulting in the overproduction and excretion of metabolites. Further progress should, therefore, not be too difficult.

10. Economic Considerations

There is still no reliable basis for the calculation of prices. In connection with the production of protein from molecular hydrogen and carbon dioxide, an "electrical biomass equivalent" has been calculated (Schlegel, 1969). It is based on the experimental data with respect to the amount of hydrogen consumed and biomass produced. The hydrogen which can theoretically be produced from 1 kWh by electrolysis of water is equivalent to 66 g biomass or at least 33 g bacterial protein. If, for comparative purposes, one considers the costs of other unconventional crude materials such as hydrocarbons or methane and includes costs for oxygen, carbon dioxide and mineral salts, biomass production on the basis of hydrogen can compete successfully with the processes mentioned.

The costs of the carbon and energy sources have only a minor influence on the price of the final protein product; the price is mainly determined by the manufacturing costs. Even when the substrate may be obtained free, as in the case of sulfite liquor or waste whey, the price of producing the fodder yeast is still considerable (Humphrey, 1967; Tannenbaum and Mateles, 1968).

In our opinion, the quality of the products will be the deciding factor in determining which method will be given preference.

11. Conclusions and Summary

The hydrogen bacteria are well suited for the transformation of the energy of molecular hydrogen to organic material. As far as their growth rate, vigour, biosynthetic efficiency, and yield are concerned they fulfil all demands. However, no conclusions can be drawn from the investigations on the nutritive value of hydrogen bacteria until more bacterial strains have been tested for a greater number of parameters, such as biological value of the proteins and toxicity of accompanying cellular substances. Since the gaseous nutritional components, molecular hydrogen, oxygen, and carbon dioxide can easily be supplied as rather pure substances, the danger of contamination of the biomass by exogenous toxic (cancerogenous) materials is minimal. From this viewpoint the hydrogen bacteria are directly comparable to green algae and plants. The hydrogen bacteria are accessible to genetic manipulations and eventually may be used for the production of metabolites.

References

Blackkolb, F., Schlegel, H. G.: Arch. Mikrobiol. **62**, 129—143 (1968a).
— — Arch. Mikrobiol. **63**, 177—196 (1968b).
Brown, L. R., Cook, D. W., Tischer, R. G.: Developm. Industr. Microbiol. **6**, 223—228 (1964).
Calloway, D. H., Kumar, A. M.: Appl. Microbiol. **17**, 176—178 (1969).
Chapman, D. D., Meyer, R., Proctor, C. M.: Developm. Industr. Microbiol. **4**, 343—354 (1963).
Dicks, J. W., Tempest, D. W.: J. gen. Microbiol. **45**, 547—557 (1966).
Eberhardt, U.: Bakt. I, Suppl. 1, 155—169 (1965).
Foster, J. F., Litchfield, J. H.: Biotechnol. Bioeng. **6**, 441—456 (1964).
Grohmann, G.: Cbl. Bakt. II **61**, 256—271 (1924).
Herbert, D.: Symp. Soc. Gen. Microbiol. **11**, 391—416 (1961).
Hill, F., Schlegel, H. G.: Arch. Mikrobiol. **68**, 1—17 (1969a).
— — Arch. Mikrobiol. **68**, 18—31 (1969b).
Humphrey, A. E.: Biotechnol. Bioeng. **9**, 3—24 (1967).
Just, F., Schnabel, W.: Branntweinwirtsch. **2**, 113—115 (1948).
— — Die Brauerei, wiss. Beilage, **4**, 57—60; **5**, 8—12 (1951, 1952).
Kistner, A.: Proc. Koninkl. Ned. Akad. Wetenschap., Ser. C **56**, 443—450 (1953).
— Proc. Koninkl. Ned. Akad. Wetenschap., Ser. C **57**, 186—195 (1954).
Klausmeier, R. E., Brown, L. R., Benes, E. N., Strawinsky, R. J.: Bacteriol. Proc. p. 123 (1958).
Klemme, J. H., Schlegel, H. G.: Arch. Mikrobiol. **68**, 326—354 (1969).
König, Chr., Sammler, I., Wilde, E., Schlegel, H. G.: Arch. Mikrobiol. **67**, 51—57 (1969).
Krebs, H. A.: Biochem. J. **80**, 225—233 (1961).
Mateles, R. I., Baruah, J. N., Tannenbaum, S. R.: Science **157**, 1322—1323 (1967).
— Tannenbaum, S. R.: Cambridge: The M.I.T. Press 1968.
McGee, J. M., Brown, L. R., Tischer, R. G.: Nature **214**, 715—716 (1967).
Nilsson, E., Jönsson, A. G.: Kgl. Lantbrukshögsk. Ann. **28**, 203—214 (1962).
Ogur, M.: Developm. Industr. Microb. **7**, 216—220 (1966).
Reh, M.: 2. Symp. Techn. Mikrobiol. (1970; in the press).
— Schlegel, H. G.: Arch. Mikrobiol. **67**, 99—109 (1969a).
— — Arch. Mikrobiol. **67**, 110—127 (1969b).

Roberts, L. P.: Nature **172**, 351 (1953).
Rudolph, V.: Göttingen: Thesis 1968.
Ruhland, W.: Ber. Dtsch. Bot. Ges. **40**, 180—184 (1922).
— Jb. wiss. Bot. **63**, 321—389 (1924).
Schlegel, H. G.: Raketentechn. Raumfahrtforschg. **8**, 65—67 (1964).
— In: Fermentation Advances. Perlman, D. (Ed.), p. 807—832. New York: Academic Press Inc. 1969.
— Hill, F. F.: Pharma Int. **5**, 189—198 (1969).
— Kaltwasser, H., Gottschalk, G.: Arch. Mikrobiol. **38**, 209—222 (1961).
— Lafferty, R. M.: Zbl. Bakt. 2. Abt. **118**, 483—490 (1964).
— — Nature **205**, 308—309 (1965).
— — Krauss, I.: Arch. Mikrobiol. **71**, 283—294 (1970).
— Schuster, E., König, Chr.: Zbl. Bakt. 1. Abt. Orig. Suppl. **2**, 73—78 (1967).
— — Reh, M., Metz, H.: Zbl. Bakt. 2. Abt. **119**, 225—231 (1965).
Schuster, E.: Göttingen: Thesis 1967.
— Schlegel, H. G.: Arch. Mikrobiol. **58**, 380—409 (1967).
Tannenbaum, S. R., Mateles, R. I.: Science J. **4**, 87—92 (1968).
— — Capco, G. R.: In: Adv. Chem. Series **57**, World Protein Resources. Gould, R. F. (Ed.), p. 254—268. Am. Chem. Soc., Washington 1966.
Tsao, G. T.: Biotechnol. Bioeng. **10**, 765—785 (1968).
Waslien, C. I., Calloway, D. H.: COSPAR Conference Abstr., p. 163 (1967).
— — Appl. Microbiol. **18**, 152—155 (1969).
— — Margen, S.: Nature **221**, 84—85 (1969).
Wilde, E.: Göttingen: Thesis 1961.
— Arch. Mikrobiol. **43**, 109—137 (1962).

H. G. Schlegel and R. M. Lafferty
Institut für Mikrobiologie der Universität Göttingen
und Institut für Mikrobiologie
der Gesellschaft für Strahlenforschung mbH, München
3400 Göttingen, Grisebachstraße 8

B. Liquid and Solid Hydrocarbons

A. Einsele and A. Fiechter

With 14 Figures

Contents

1. Introduction

Research into the production of microbial protein from hydrocarbons
has advanced in an unforeseen manner in the last 10 years. Although it
was already observed long ago (Mycowski, 1895) that micro-organisms
can use hydrocarbons as the sole source of carbon, it was only in 1954
that a proposal was made (Davis and Updegraff, 1954) that it might be
possible to produce foods from them. There are many reasons why

petroleum-microbiology was not actively developed before. Firstly, conditions do not favor the growth of organisms, since petroleum is mostly found under anaerobic conditions, while degradation proceeds only aerobically. Secondly, the extremely low water solubility of paraffins hinders intensive microbe growth under natural conditions. Hence petroleumbiology experienced no appreciable development, until petroleum prospecting was developed after the Second World War. At first, the biological de-waxing of high-boiling distillates was considered the most promising utilization of the ability of micro-organisms to assimilate hydrocarbons.

Example

Jet planes flying at high altitudes require kerosene with solidifying points below $-50°$ C. Biological de-waxing of a distillate having a boiling point between 170° C and 200° C, can induce the required property with the simultaneous production of nutrient protein.

It was only later that the resulting bio-mass was tested for other purposes. In recent years, petroleum companies have started using hydrocarbon distillates as raw material for protein synthesis. Thus, new prospects have emerged for industrial protein production. These modern possibilities of synthesis are mentioned repeatedly in responsible quarters, when the possibilities of a worldwide campaign against hunger are discussed.

This new branch of petrochemistry is very important, not only in protein production, but also in the area of biosynthesis of other compounds. From this we may conclude that the interest of the petrochemical industry in these biological processes does not necessarily stem from an enthusiasm for combating hunger in the world.

2. Microbial Degradation of Hydrocarbons

a) Aliphatic Substrates

The degradation of hydrocarbons by micro-organisms occurs quite specifically. Zobell (1946) tried to formulate this specificity by means of four rules:

1. Aliphatic compounds are more readily attacked by micro-organisms than aromatic compounds.

2. Long chains are degraded preferentially as compared with short chains.

3. Unsaturated compounds are degraded more readily than saturated compounds.

4. Branched chains are degraded more readily than straight chains (see Section 2b).

There are examples which show that these rules are not generally true. Still, they show the specificity of the degradation.

Rule 1, however, has unrestricted applicability. The degradation of aromatic compounds proceeds in a much more complicated manner and cyclic compounds offer great resistance to oxidation by microorganisms.

Rule 2. The length of the aliphatic carbon chain is decisive for the biological degradation. Most organisms show very good growth on higher alkanes with a chain of more than 18 C atoms. Except for some *Pseudomonas* strains, which can use hexane, heptane and octane as substrate, chains with 8 C atoms are the shortest alkanes which can still be assimilated. Low-molecular, liquid alkanes (with less than 8 C atoms) which are able to dissolve lipids act toxically on many organisms. A few grow only if the concentration of low-molecular hydrocarbons is below the saturation constant. Possibly, this toxicity is due to the fact that the phospolipid micelles of the cell membrane are destroyed by low-boiling alkanes.

On the example of *Candida lipolytica*, Figs. 1 and 2 show a typical picture for the assimilability of alkanes (Munk *et al.*, 1969).

This yeast strain has the capacity of assimilating (oxidizing) alkanes and also of producing biomass, only if the substrate has an aliphatic chain with at least 8 C atoms. A single C atom more or less can therefore decide whether the substrate can be utilized or not.

The higher homologs above this critical chain length are all assimilable. However, the assimilations differ in biomass formation and CO_2 production. These substrates are more or less completely oxidized to CO_2.

Data on assimilation rates were obtained by experiments with alkane mixtures (Dostalek *et al.*, 1968; Glikmans, 1968). Yeasts do not simultaneously degrade all the alkanes in mixtures. In an initial phase, short-chain compounds are degraded more rapidly than longer-chain compounds. Accordingly, Dostalek classified the assimilable alkanes into three groups:

first group: $C_{12}–C_{14}$,
second group: $C_{15}–C_{17}$,
third group: C_{18} and higher.

At first, the low-molecular compounds are degraded fastest. Glikmans and Dostalek concurrently found that the assimilation rate decreases with increasing molecular weight. This behavior seems to be due to differences in the degradation kinetics (selectivity of cell wall, specific reactivity of the alkanes) which will be discussed later.

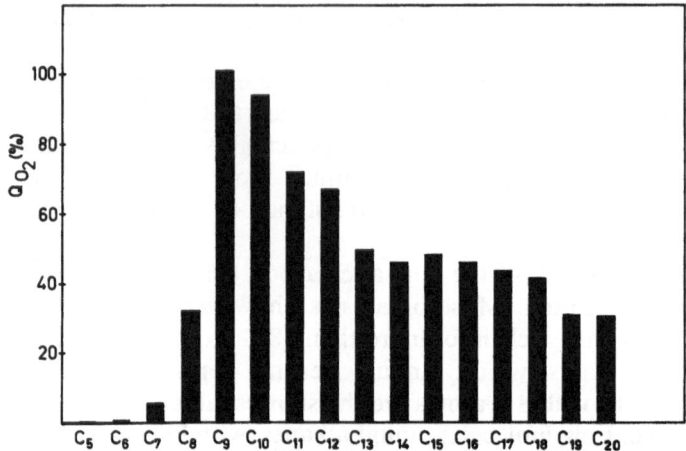

Fig. 1. Oxidation of pure *n*-alkanes by *Candida lipolytica 4-1*. Q_{O_2}: Specific oxygen uptake after deducting endogenous respiration; oxidation of nonane = 100%, C_5–C_{20}: Individual hydrocarbons denoted by number of carbons in molecule (Munk *et al.*, 1969)

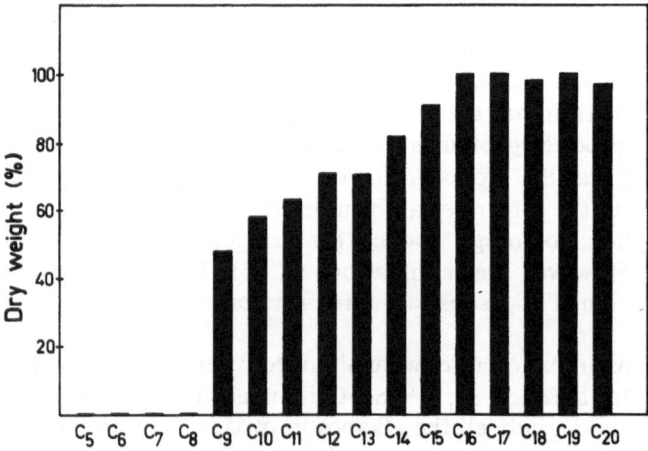

Fig. 2. Growth of *Candida lipolytica 4-1*. Dry weight: Biomass after 70 hours' cultivation, expressed in percentage form; growth on hexadecane = 100%. C_5–C_{20}: Individual hydrocarbons, denoted by number of carbons in molecule (Munk *et al.*, 1969)

Zobell's *Rule 3*, according to which unsaturated compounds are more easily degraded than saturated compounds, has been contradicted by so many experiments that the reverse formulation seems more in keeping with the facts.

In particular, unsaturated alkanes lead to much lower yields as compared with the corresponding saturated homologs.

b) Branched and Cyclic Substrates

The primary steps in the oxidation of iso- and cyclo-alkanes are probably the same as in the case of *n*-alkanes. However, the number, size and arrangement of side chains at the alkane molecule appear to have a stronger influence on the oxidation capacity. No general rules have yet been worked out, and only a few points are clear:

Alkanes with short side chains (methyl groups) are oxidized, but much more slowly. The number of methyl groups therefore influences the oxidation, i.e. Zobell's fourth rule is proved. Alkyl-aromatic hydrocarbons with a side chain of less than 5 C atoms in the molecule are toxic for yeast cells.

Biomass production depends on the chain length and also on the position at which the side chain is attacked. Iso-alkanes which are branched at both ends are generally not assimilated (e.g. 2-13-dimethyl tetradecane). In this case the point of attachment is blocked on both sides.

Compound	% Destroyed
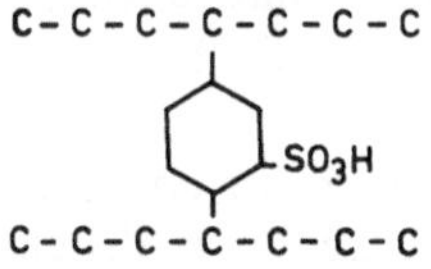 C–C–C–C–C–C–C / C–C–C–C–C–C–C, ring with SO₃H ($C-C-C-C-C-C-C$ ring SO_3H)	17
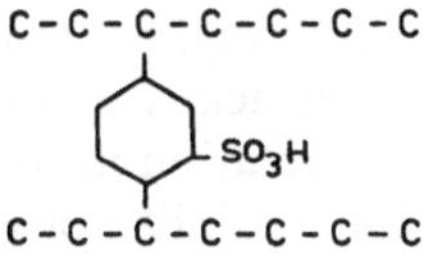 C–C–C–C–C–C–C / C–C–C–C–C–C, ring with SO₃H	40
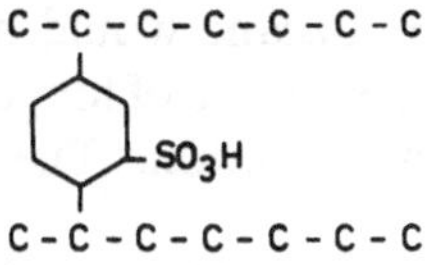 C–C–C–C–C–C–C / C–C–C–C–C–C–C, ring with SO₃H	80
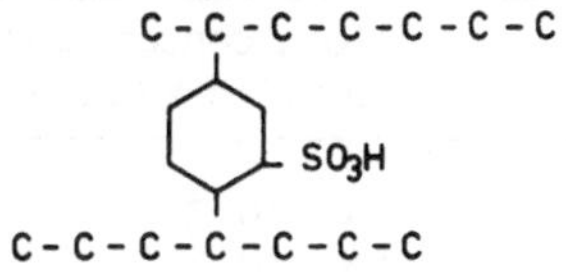 C–C–C–C–C–C–C / C–C–C–C–C–C–C, ring with SO₃H	45

Fig. 3. Degradation of diheptylbenzene sulphonate by 24-hour activated sludge treatment (Swisher, 1963)

Iso-alkanes with a branched end give less than 50% biomass as compared with the corresponding *n*-alkane. Thus, only about 20% biomass is formed from 8-methyl-pentadecane, as compared with the production on hexadecane as substrate with the same molecular weight.

The conditions with diheptyl-benzene-sulphonate (Swisher, 1963) show particularly clearly how greatly the length of the unbranched chain-end can influence degradation.

In this experiment with a mixed culture of *Pseudomonas* species, the unbranched part of the chain must be at least 4 C atoms long. The sulphonated diheptyl chains are more readily obtained if the free ends are longer. The maximum degradation of 80% occurs only when the unbranched ends of the two chains have a length of 5 C atoms.

c) Substrate Solubility

The solubility of *n*-alkanes in water is very low and decreases rapidly with increasing number of C atoms. McAuliffe (1963) experimentally determined the solubilities up to 8 C atoms. Johnson (1964) calculated further values by extrapolation, and for chain lengths of more than 8 C atoms he found such low values that, according to conventional concepts, the uptake of such compounds by the living cell becomes critical.

Accordingly, tetradecane has a saturation constant of 9.8×10^{-10} mol; here, we have to consider that at 25° C, hydrocarbon chains up to 18 C atoms are still liquid. Hopkins and Chibnall (1932) even observed the growth of *Aspergillus versicolor* on straight chains with 34 C atoms. Such compounds should therefore be attacked in solid form. However, an objection has been raised that extrapolation in the manner described leads to wrong solubility values. The solubility of C_{10}–C_{18} molecules does not decrease as strongly as was calculated by Johnson by extrapolation. According to Baker (1956), decane would have a solubility of 16 p.p.m., and octadecane 6 p.p.m. The uptake of longer chains is thought to contribute to this solubility stabilization. The problems of low

Table 1. *Solubility of n-alkanes in water at 25° C*

Nature of *n*-alkanes	Molar concentration of saturated solution
Hexane	1.1×10^{-4}
Octane	5.8×10^{-6}
Decane	3.1×10^{-7}
Dodecane	1.7×10^{-8}
Tetradecane	9.8×10^{-10}

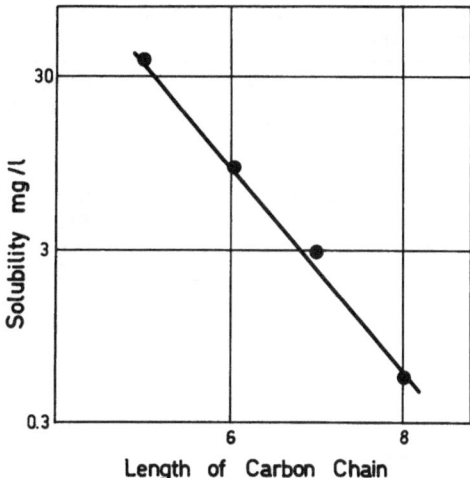

Fig. 4. Solubility of *n*-alkanes in water (McAuliffe, 1963)

solubilities finally amount to the question of whether the substrate transfer proceeds from the water phase to the cell, or directly from the alkane phase by direct contact with the cell.

d) Media

Although the mass balance for balanced growth as compared with carbohydrates is not appreciably different, the ratio of nitrogen addition to the C component (C/N ratio) and the form in which N is added must be specially noted. Table 2 and Fig. 5 show a comparison of the various possibilities. From this we see that N is best added as $(NH_4)^+$: on the one hand, it reduces the generation time, and on the other, it increases the yield.

With $(NH_4)_2 SO_4$ as N source, the addition of yeast extract causes a significant reduction in the generation time from 2.1 to 0.5 h, but no effect is observed in the case of urea + yeast extract. The energy conditions are so modified as compared with carbohydrates, that it also seems necessary to check the phosphorus added. Noyes (1969) has formulated a limit for the mineral salt additions.

The culture conditions play a large part in the formation of the product. The greater the C/N ratio, the more fat is produced (Ratledge, 1968). On the other hand, the production of keto acids decreases with increasing C/N ratio.

In the same paper, it is also pointed out that there are no special requirements regarding vitamin additions to the medium.

Table 2. *Growth rate and cell yield of cultures grown on various media with n-alkanes* C_{14}–C_{17} *as carbon source (Wagner et al., 1969)*

		Generation time, hr	Conc. of cells, g/L	Y
Nocardia NBZ 23	A	0.8	14.7	98.0
	B	2.1	15.3	102.0
	C	0.5	20.5	111.0
	D	4.0	14.0	93.5
	E	4.0	15.7	85.0

$A = NH_4NO_3$. — $B = (NH_4)_2SO_4$. — $C = B +$ yeast extract. — $D =$ Urea. — $E = D +$ yeast extract.

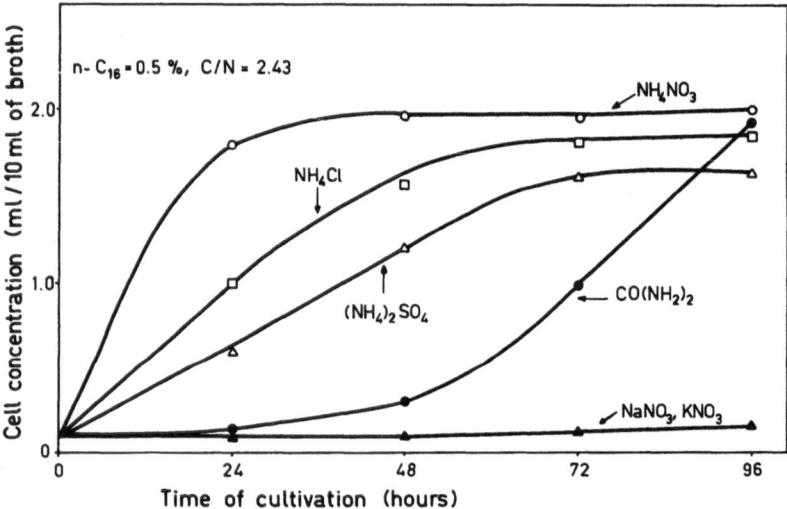

Fig. 5. Effects of nitrogen sources on the cell yield of *Candida tropicalis* N_7Y_1 (Yamada *et al.*, 1968)

Table 3. *Composition of media (Noyes, 1969)*

Mineral		Preferred weight percent of salts in aqueous medium, based on 1 weight percent concentration of growing cells
H_3PO_4	P	0.01 —1.0
Na_2SO_4	S and Na	0.01 —0.5
KCl	K and Cl	0.01 —0.5
$MgSO_4$	Mg	0.005—0.5
$CaCl_2$	Ca	0.005—0.5
$FeSO_4$	Fe	0.001—0.1
$MnSO_4$	Mn	0.001—0.1

e) Organisms

The capacity to assimilate alkane is very widespread in micro-organisms. Most of the yeasts, bacteria and mould fungi appear to be able to utilize hydrocarbons, though very poorly in some cases. Many alkane-assimilating strains have been isolated and identified from oilfields, tank plants and oil-contaminated waters. Some important and extensive papers of earlier and recent times show the complexity of the micro-organisms which can assimilate hydrocarbons (Fuhs, 1961; Iizuka et al., 1966). Fuhs isolated about 100 bacteria strains. In this report he noted only 5 yeast strains, thus confirming the earlier assumption that alkane-degradation was due to bacterial cells. Klug et al. (1967), however, isolated over 100 Candida strains, some of which showed very good growth on alkanes. Hence the yeasts seem to move into the foreground.

If we survey the bacteria which can assimilate alkanes, the picture is not taxonomically uniform. Best represented are the families *Pseudo-monaceae*, *Enterobacteriaceae*, *Micrococcaceae* and *Bacillaceae*. A particularly large number of investigations have been carried out with strains of *Pseudomonas* and *Mycobacterium*.

Regarding the yeasts, the picture is clearer since the family of *Cryptococcaceae*, belonging to the *Fungi imperfecti*, is primarily known as a utilizer of hydrocarbons. The extent to which these capacities can also be attributed to the *Saccharomycetae* is disputed. For example, if, in agar surface cultures, the substrate is offered to the organisms only in the form of individual alkane molecules, the *Saccharomycetae* also show growth. Minimal amounts already have a toxic effect on the cells and submerged culture is impossible. The obvious assumption is that *Cryptococcaceae* have cell-wall structures which are resistant to the influence of solvents (alkanes). So far, only the fodder yeasts (*Candida* sp.) have any economic importance.

Many investigations were also conducted with mould fungi (Nyns et al., 1968). Some strains of *Aspergillus*, *Penicillium*, *Fusarium* and *Cladosporium* grow on hydrocarbon mixtures (C_{12}–C_{14}). Mould fungi can, however, assimilate short-chain alkanes in extremely few cases (C_{16}–C_{11}).

3. Metabolism

a) Transport

As already mentioned, transport into the cell is a special problem, because of the slight solubility of the substrate. Probably the alkane molecules penetrate into the cell, either from the hydrocarbon phase or dissolved in water. It was shown with *Candida lipolytica* that the hydro-

carbon molecules penetrate the cell wall and accumulate on the cytoplasmic membrane (Ludvik *et al.*, 1968). Yeasts grown on hydrocarbon
show morphological variations as compared with those grown on
carbohydrates. Where hydrocarbons have penetrated, the cytoplasmic
membrane is thicker and shows deep invaginations. This membrane is
involved in the metabolism and also in the transport of alkanes. The
position at which the first enzymatic oxidations occur has, however, not
been determined. This may occur at the cytoplasmic membrane, or the
alkanes may diffuse passively into the cytoplasm through the assumed
pinocysts in the invaginations. In any case, the plasma membrane
plays a decisive role in the penetration of hydrocarbons. It is responsible
for the transfer of substance into and from the cell. Some enzymes
involved in this are localized on the outer side of the plasma membrane,
and probably also those of the first oxidation step (Ludvik *et al.*, 1968).
It must be assumed that the alkane molecules penetrate the cell wall
as a whole through passive sorption (diffusion). Only the subsequent
transport through the plasmatic membrane is an active metabolic process.

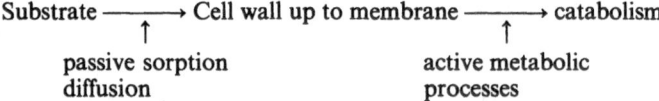

Possibly the active transport takes place in acid form. Fatty acids penetrate better when their chain length is shorter (Suomalainen, 1968).
Thus, the faster assimilation rate of short-chain alkanes would be

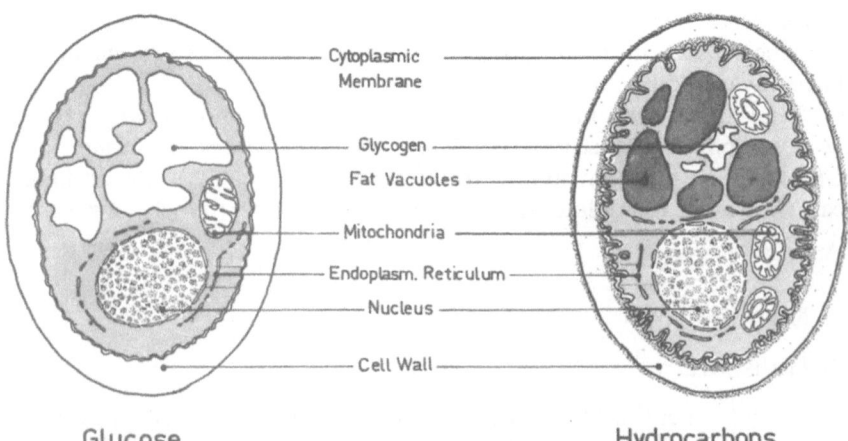

Fig. 6. Schematic picture of an ultrathin section of *Candida lipolytica* grown on
glucose or hydrocarbon medium (Ludvik *et al.*, 1968)

explained. The plasma membrane contains lipases; the study of the lipid structure of this membrane could give indications about the selective regulation of the penetration of components into and from the cell.

b) Metabolic Mechanisms

Many points about the mechanism of hydrocarbon degradation are still not clear. Possible ways have been suggested, but only a few of the postulated intermediate products have been found. Generally, degradation proceeds mainly from alkane via primary alcohol and aldehyde up to the corresponding fatty acid. Further, the fatty acid formed is degraded via β-oxidation, or again oxidized to the dicarboxylic acid, and then de-carboxylation takes place. Fig. 7 illustrates the most important ways of degradation (McKenna and Kallio, 1965).

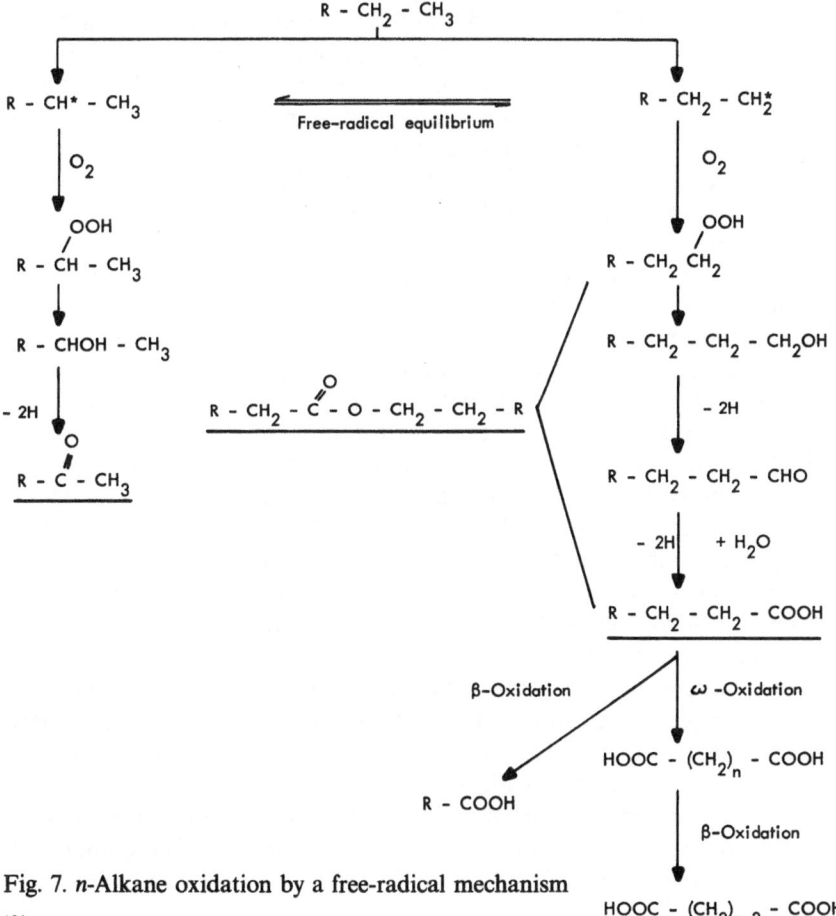

Fig. 7. *n*-Alkane oxidation by a free-radical mechanism

12*

Direct incorporation of molecular oxygen as $^{18}O_2$ (Imada et al., 1967) shows that the initial reaction is catalysed by an oxygenase and necessarily depends on oxygen. The most favorable position for attachment is the C_1 position. The only stable metabolic product which could be isolated is the homologous normal alcohol. Besides this so-called C_1 oxidation, there exists as another possibility, C_2 oxidation. It could be shown from the example of Pseudomonas methanica that propane is oxidized into a mixture of propionic acid and acetone. In this case, n-alkanes are converted by intact cells into methyl ketone and also into the corresponding fatty acid. Other assumptions postulate that the C_2 oxidation occurs on the basis of an equilibrium between the radicals of the C_1 and C_2 positions.

$$RCH_2CH_2^* \rightleftharpoons RCH^*-CH_3 .$$

The reduction of the methyl ketones produced gives secondary alcohols.

Azoulay et al. (1963) reported that the first step of alkane degradation takes place with the help of a dehydrogenase, i.e. that oxygen is not necessarily required. But, by now, this possibility is considered quite improbable. Moreover, a di-terminal oxidation of alkanes appears to exist (Iizuka et al., 1966). In this case the hydrocarbon chain is oxidatively attacked on both sides, and dicarboxylic acids are formed. This is important because it could thus be shown that the degradation paths finally open into the TCA (as proved by the formation of succinic acid).

Olefines are used as substrates by a large number of micro-organisms. Relationships in metabolism with n-paraffins are suggested by the similarity of the C skeleton. The earlier assumption that olefins are more suitable for growth was found to be wrong. Cultures with Pseudomonas sp. which grow with 1-heptene showed that the most important degradation path begins with the oxidation at the saturated end of the molecule. The most complicated metabolic conditions are found in the case of the aromatics. There are detailed descriptions of this in the literature (van der Linden et al., 1965). Fig. 8 gives an example of ring opening in aromatics. The ring is first oxidized in a perhydroxylation step. This produces a trans-dihydro-dihydroxy compound, which is then dehydrogenated to the dihydroxy compound. Ring cleavage takes place through the action of an oxygenase. In the case of a single-ring aromatic compound, the usual intermediate is a cis-cis muconic acid. The muconic acid forms a lactone which undergoes isomerization to a cis-trans isomeride. This is further oxidized to α-ketoadipate which is shunted into the tricarbocylic acid cycle via acetyl-CoA and succinate. With regard to multiple-ring systems, the dihydroxy compound is usually oxidized to a 1-hydroxy-2-oxobut-3-enoic acid. The compound can readily be decarboxylated and oxidized to give a compound which

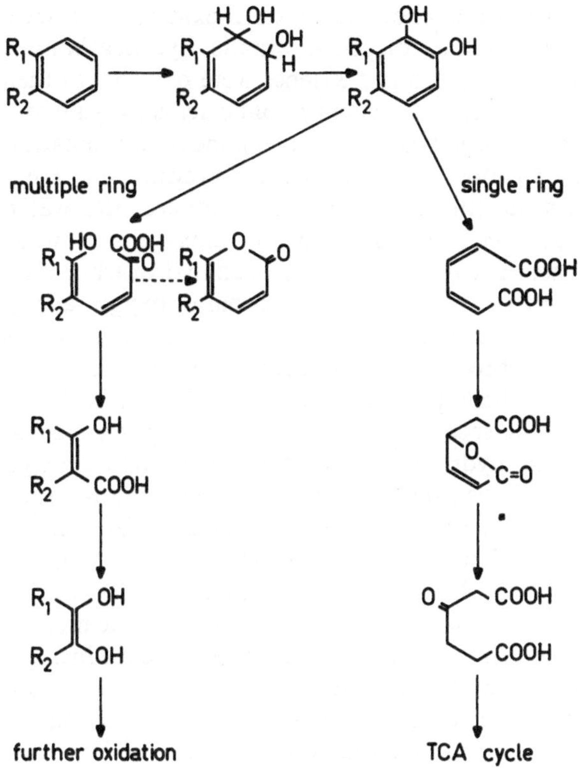

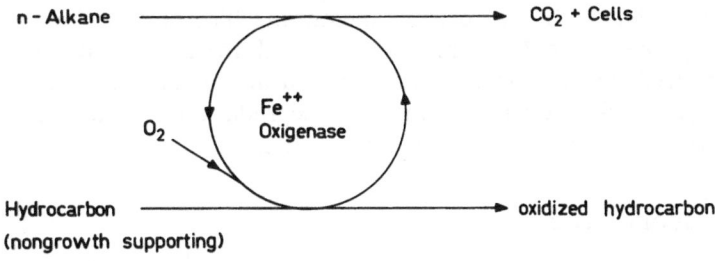

Fig. 8. General mode of ring splitting (Humphrey, 1967)

Fig. 9. Co-oxidation of hydrocarbons

may exist in various tautomeric forms depending upon the pH of the solution.

Co-oxidation plays a special role in connection with hydrocarbons (see Fig. 9) (Raymond *et al.*, 1967, 1969). Aromatics and cyclic paraffins are normally good examples of Co-oxidation. Nocardia cultures have been shown to bring about significant oxidation of several methyl-substituted mono- and dicyclic-aromatic hydrocarbons. Under Co-oxidation conditions, o- and p-xylenes were oxidized to their respective monoaromatic acids, o-toluic and p-toluic acids (Raymond, 1967).

Normally, microbial oxidations of alkanes and aromatic components are catalysed by mono-oxidases which introduce one atom of oxygen into the substrate (the second atom is reduced and water is formed, along with an electron donor), or by dioxygenases which add two atoms of oxygen. The initial oxidation of all hydrocarbons is oxygenative. Alkane hydroxylases incorporate one atom of oxygen and give primary alcohols. Electrons are transferred from a reduced nicotinamide nucleotide to rubredoxin, a nonheme iron protein in *Pseudomonas oleovans* or to a hemoprotein (cytochrome P-450) in *Corynebacterium*, which reduces the second atom of oxygen into water during hydroxylation. The alkane hydroxylases and the alcohol and aldehyde dehydrogenases act on substrates of different chain lengths. But, with some exceptions, it could be shown that the capacity for oxidizing paraffins can be induced (Nyns *et al.*, 1969; Ribbons, 1969). With *Pseudomonas*, it could be proved that the substrate induces the synthesis of specific oxidation enzymes and not a proper transport system (van Eyk, 1968). *Candida lipolytica* has a constitutive enzyme system which can assimilate acetyl alcohol and palmitic acid. Likewise, the enzymes of alkane oxidation, which give the alcohol (primary oxidation step) can be induced. The alkane uptake can be split up in time into two phases: the first is as long in cells with an induced enzyme system as it is in those with a non-induced system and lasts 1–2 minutes. This corresponds to the time required by the molecule to penetrate into the cell wall and possibly as far as the cytoplasmic membrane. The second phase covers active uptake and is therefore associated with metabolic processes. In a non-induced system, for example, *Candida intermedia* assimilates octadecane after 3 h, while an induced degradation system begins to decompose the substrate already after 5–10 min.

c) Metabolic Products

Besides the production of biomass, the investigators are also trying to produce interesting intermediate products. Thus, the production of fatty acids has been intensively investigated by the preparation of definite fatty-acid samples. Yeasts grown on media which do not contain hydrocarbons contain fatty acids whose C chains are practically all even-numbered. Quantitatively detectable differences in the fatty acid sample can be attributed to nutrient combination which obviously exerts a decisive influence.

The relative proportions of the fatty acids formed correspond, in the case of hydrocarbons, to the chain-length of the substrate. If even-numbered chain lengths are used, we likewise obtain acids with an even number of C atoms; the acid with the same number of C atoms as the substrate predominates. The same applies to odd-numbered chain lengths, the mechanism of whose formation has, however, not been explained. Since normal degradation leads to C_2 units and re-synthesis leads to even-numbered chains, a modification of the anabolic processes must be present. The even-numbered fatty acids from odd-numbered alkanes are formed by *de novo* synthesis (Makula *et al.*, 1968; Dunlap *et al.*, 1967). This mechanism does not yield fatty acids with an odd number of C atoms, as shown in Table 4.

Table 4. *Percentage of fatty acid composition from even-carbon n-alkane oxidations*[a] *(Makula et al., 1968)*

Compound	Decane	Dodecane	Tetradecane	Hexadecane	Octadecane
C 10:0	3.78	0.46	1.70	0.82	—[b]
C 11:0	—[c]	—	—	—	—
C 12:0	7.55	6.65	7.25	7.14	8.92
C 13:0	—	—	—	—	—
C 14:0	4.36	0.84	28.43	2.04	1.93
C 15:0	—	—	—	—	—
C 15:1	—	—	—	—	—
C 16:0	39.50	23.26	10.88	29.58	16.34
C 16:1	14.52	14.30	16.93	43.34	20.80
C 17:0	—	—	—	—	—
C 17:1	—	—	—	—	—
C 18:0	6.68	6.63	11.60	7.14	10.40
C 18:1	23.61	47.52	23.21	2.04	41.60

[a] Fatty acid composition is expressed as a percentage of total fatty acids.
[b] Trace amounts.
[c] Not detectable.

Intermediate products (e.g. fatty acids) can have a toxic effect (Fencl, 1961). As a provisional explanation, we assume that the dissociated forms of the fatty acids no longer guarantee the passage of anions. The enhancement of assimilation by dialysis culture and specific growth inhibition by its dialyzable material, as noticed in a kerosene medium, were demonstrated also in the case of individual *n*-alkane, i.e. hexadecane or decane. These findings seem to indicate that the effect of dialysis culture on alkane assimilation is a fairly general phenomenon. From the evidence presented for the growth inhibition by the dialysable substance, it is likely that the improvement of hydrocarbon assimilation by dialysis culture may be attributed to the presence of "growth-inhibiting factor" in the dialysable material (Aiba *et al.*, 1969). Gradova *et al.* (1968) showed that in chemostatic experiments with yeasts in media with *n*-alkanes under optimum conditions, the most important metabolic products are never enriched. Table 5 gives a survey of possible new products which can be formed during alkane oxidation.

Table 5. *Metabolic products*

Product	Organism	Substrate	Author
L-Glutamic acid	*Corynebact. hydrocarboclastus* S 10 B 1	Dibasic carboxylic acid	Ikeda *et al.*, 1969
Cytochrome C	*Cand. albicans*	*n*-Hexadecane	Tanaka *et al.*, 1967a and b
Carotenoids	*Mycobacterium smegmatis*	*n*-Alkane-mixture	Tanaka *et al.*, 1968
Coenzyme Q	*Candida tropicalis*	*n*-Alkane-mixture $(C_{12}–C_{14})$	Shimizu *et al.*, 1969
Salicyclic acid	*Pseudomonas aeruginosa*	Naphtalene	Wodzinski, 1968
Lactic acid	*Arthrobacter oxydans*	1-2-Propanediol	Yagi *et al.*, 1969
L-Tryptophan	*Bacterium*	Hydrocarbon	Kyowa Hakko Kogyo, 1969
L-Citrullin	*Corynebacterium*	*n*-Alkanes	Tokoro *et al.*, 1969
Vitamins Group B	*Candida*	*n*-Alkanes	Popova, 1968
Antibiotic pyrrolnitrin	*Pseudomonas multivorans*	Toluene	Mabe *et al.*, 1970

d) Role of Oxygen

Assuming that the yield is 100% with respect to hydrocarbon and 50% with respect to glucose, the process with alkanes requires 2.6 times more oxygen for the production of the same quantity of cells. In the microbial oxidation of hydrocarbons, cases of limited oxygen supply have been reported (Klug and Markovetz, 1969; Yamada *et al.*, 1968).

Figs. 10 and 11 show clearly that insufficient aeration reduces yield. Similarly, growth rates are positively influenced by an adequate oxygen supply, so that in continuous culture, the productivity $(D \cdot X)$ is strictly

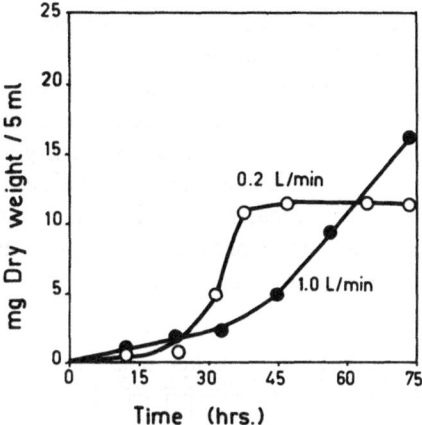

Fig. 10. Effect of aeration on growth response (Klug *et al.*, 1969)

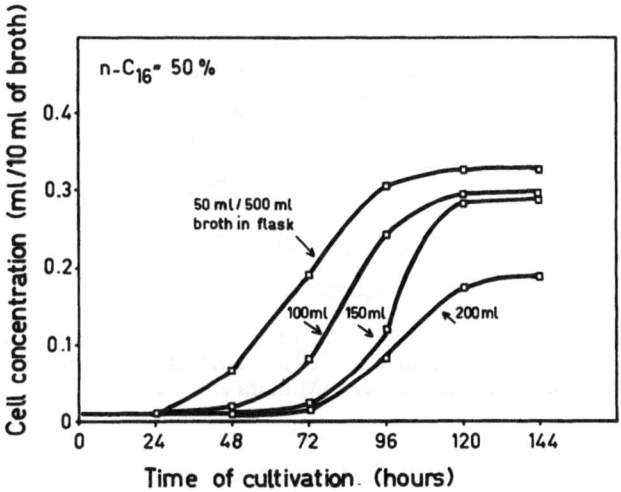

Fig. 11. Effect of aeration on the cell yield of *Candida tropicalis* N_7Y_1 (Yamada *et al.*, 1968)

controlled by the oxygen transfer. For such reasons, Glikmans (1968) found that the highest productivities were only $0.88 \, g \cdot l^{-1} \, h^{-1}$. A complete understanding of oxygen transfer is rendered difficult by the fact that the solubility of oxygen in the oil phase is greater than in the water phase, and the interfacial conditions between air and water are disturbed in the presence of paraffins by changes in the surface tension (Calderbank, 1967). On adding paraffins to aqueous sulfite solutions, we get much higher oxygen transfer rates (cf. Fig. 12); this is due to the more favorable interfacial conditions in the oil-water system, as well as the greater actual oxygen concentration.

The existing three-phase system (oil-water-air) therefore complicates the measurements of oxygen transfer. Methods of determining oxygen with aqueous sulfite solutions as the model system give values which are characteristic for the apparatus. While sulfite values have sometimes been questioned as a suitable means of prediction of biological experiments, our own experiments have proved their applicability to growth in media containing hydrocarbons. In bio-reactors which attained

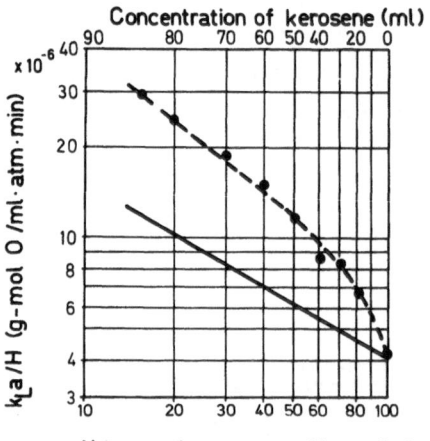

Fig. 12. Effect of kerosene concentration on $k_L a/H$. Total volume of liquid; kerosene and aqueous sulfite solution = 100 ml; straight line shows the oxygen transfer rate coefficient without kerosene (Mimura *et al.*, 1969)

sulfite values up to 1200 mMol $O_2/l \cdot h$, it could be shown that oxygen is not the limiting factor. As a result of better mixing and stirring devices, the growth rate with *Candida lipolytica* on hexadecane as substrate could be increased from $0.22\,h^{-1}$ to $0.31\,h^{-1}$ (Einsele, unpublished).

Objections can also be raised against the use of polarographic methods to determine the solubility of oxygen. Since the solubility can be simultaneously measured only in one phase, polarography is categorically not applicable in liquids with two phases, as is the case with hydrocarbon fermentation. Such measurements with oxygen electrodes were conducted by Buswell and Jurtshuk (1969). A direct influence of oxygen supply can be observed in the composition of the biomass produced. Intensive aeration causes an increase in saturated fatty acids with an even number of C atoms and a decrease in unsaturated fatty acids. The interpretation is that, with slight aeration, the fatty acids are directly incorporated after the first oxidation. With intensive aeration, there is increased oxidation into short fragments, followed by re-synthesis.

4. Cultivation

a) Emulsification

Substrate transfer plays a very special role in the three-phase system of hydrocarbon fermentation. In principle, there are two ways in which the substrate may have access to the organism.

1. Alkanes first dissolve in the H_2O phase and from there they reach the cell.

2. They are transferred direct from the hydrocarbon phase to the cell.

The organisms living in the aqueous phase therefore have a choice of taking up the dissolved substrate, or of direct absorption from the substrate phase itself. As already mentioned, the solubility of alkanes in water is so low that growth, if based only on the water-soluble substrate, would take place extremely slowly because of the low rate of substrate transfer. Obviously an additional transfer takes place directly from the hydrocarbon phase. But this implies that the cells will be adsorbed on the emulsified hydrocarbon droplets. It will therefore be favorable to make the interface area, i.e. the surface of the hydrocarbon phase, as large as possible by a fine and stable emulsion. This should guarantee that the cells will find a sufficient area of contact surface for the transfer of the substrate required for growth. Both ways of transfer influence growth. When growth takes place primarily on the emulsified droplets, which is the case particularly for very long-chain alkanes, the size of the interface plays an important part. If the entire surface is occupied by cells, exponential growth becomes impossible. According to Erdtsieck et al. (1969), the limiting ratio substrate to cells is $0.40 \times 10^3 \ kg \ dm/m^3$ oil. The size of the emulsified droplets has, according to Bakhuis and Bos (1969), an influence on the growth rate. A degree of emulsification giving droplets of the same size as the organism cells to be cultivated provides an unfavorable configuration for substrate transfer. On the other hand, growth is good if small oil droplets adhere to the cells, or if the cells are adsorbed on larger oil droplets. In hydrocarbon culture, we have therefore to aim at the best possible emulsion in the different phases. This is obtained either by adding surface-active agents (Kobayshi et al., 1967; Aiba et al., 1969a), or by better stirring and mixing devices in bio-reactors (Humphrey, 1968; Einsele and Fiechter, 1969).

b) Growth Kinetics

It is obvious we need to set up growth models for a better understanding of this complex system. Naturally, such models are always more or less acceptable simplifications of the complicated phenomena which actually occur. Besides the views of Erickson *et al.* (1969a, b, c), which will be discussed later in detail, there are also other similar proposals (Dunn, 1968; Aiba *et al.*, 1969b). These models assume that the alkane phase is uniformly distributed in droplets of equal size, and that all organisms are likewise uniformly adsorbed on the droplets. These assumptions already represent a considerable simplification, because a poly-dispersed system is formed on mixing oil and water.

However, the conclusions from the theoretical considerations should also be considered when evaluating the experiments.

Erickson *et al.* (1969) developed a mathematical model for the description of fermentations with two liquid phases in both batch and continuous cultivation. The considerations embrace growth at the surface of the drops and in the aqueous phase. Three special cases have been examined: in the first case it is assumed that growth occurs only at the surface of the dispersed phase; in the second and third cases, growth takes place both at the interface and in the continuous phase (water). The second case assumes that the substrate equilibrium is continuously maintained between the two phases, while in the third case the consumption of substrate is limited by the transport path to the aqueous phase. Fig. 13 shows a comparison between the model and experimental data. It is assumed that growth takes place only at the

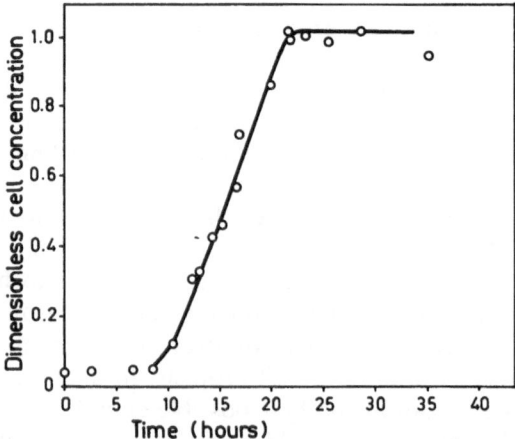

Fig. 13. Comparison of experimental data with a model which assumes that all growth occurs at the surface of the oil drops (Erickson *et al.*, 1969)

surface of the oil drops. The assumption that there is no growth in the continuous phase may be valid in this fermentation. The values of the curve thus calculated agree well with experiment.

c) Yield and Productivity

The data for yield, growth rate and productivity differ very greatly for microbial alkane oxidations. Since the growth parameters depend on the substrate and the organism, the data cannot be generalized. However, we get the impression that yield and productivity are low for short alkane chains and for cyclic chains. For medium and very long chains, typical yield values fluctuate between 70–90% (see Table 6).

Table 6. *Growth of Candida on normal alkanes (Johnson, 1965)*

Substrate	Generation time hr	Cell yield on substrate %	CO_2 produced per mole of substrate moles	Carbon recovered as cells and CO_2 %
$C_{12}H_{26}$	7.0	59.8	4.6	72.7
$C_{14}H_{30}$	6.5	83.0	5.2	84.2
$C_{16}H_{34}$	6.5	81.7	7.5	93.8
$C_{17}H_{36}$	5.0	72.9	4.3	67.6
$C_{18}H_{38}$	4.5	82.5	5.4	76.6

Temperature $= 30°$ C; pH maintained at 5.5.
The cell concentrations at the end of the growth period varied between 10 to 15 gm/litre.

The growth rate in hydrocarbon fermentations is influenced by a number of different factors and reliable values on this are rare. Fritsche (1968) found that growth rate is specific to the substrate, and is always smaller (nearly $0.2 \, h^{-1}$) than on glucose. The organism must produce additional enzymes. It must first form sufficient enzyme protein for biomass production, and this frequently leads to growth retardation. For yeasts, μ_{max} values of 0.20 to $0.44 \, h^{-1}$ were found (Azoulay, 1964). If the permeability (cf. permeability theory) and the hydrogen ion concentration (Glikmans, 1968) are optimal, it appears possible to obtain growth rates comparable with those on other substrates.

d) Separation of Biomass

Separation of the cell material obtained from hydrocarbons offers special problems. Centrifuging has to be combined with expensive solvent-extraction and additional separation methods.

The separation and drying of biomass from hydrocarbon fermentations therefore accounts for more than 50% of the production costs. Nearly one-fourth of the patents published on the microbial oxidation of hydrocarbons relate to problems of separation alone. The addition of surface-active agents means that during centrifuging the emulsion is destroyed, and we get an oily, sticky yeast mass. The complete removal of hydrocarbon residues requires multi-stage processing, during which the cells have to be treated again with surface-active agents. The remaining non-assimilated hydrocarbons must be removed as quantitatively as possible. If the biomass is to be used as a food, toxicological problems arise because residues of hydrocarbons or of the solvent used cannot be tolerated on health grounds.

A typical system for recovering cells from hydrocarbon fermentations is shown in Fig. 14.

The liquid which contains the cells and non-assimilated hydrocarbons is divided by centrifuging into three streams: cell paste, water phase and hydrocarbon phase. A part of the aqueous phase is recycled. The sediment can be treated again with O_2 and nutrient solution in a purifying fermenter and oxidized once more. Then we must centrifuge

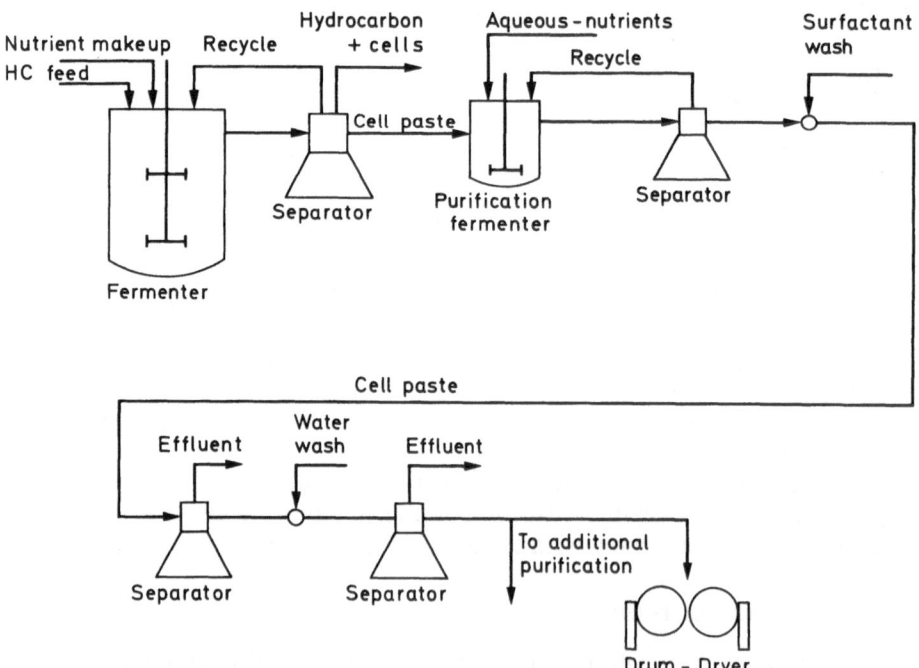

Fig. 14. Primary recovery scheme of cells from fermentation broth using hydrocarbons as substrate (Wang, 1968)

again and wash the paste. The washed cells are dried after centrifuging. The third phase of the first centrifuging, which contains the major part of the unassimilated hydrocarbons, also includes an appreciable amount of organisms. The cells of this phase can however be obtained by solvent extraction.

e) Purification

In the development of petroleum microbiology, such new aspects as the synthesis of amino acids are also receiving attention; but at present yeast production from de-waxing processes is very much in the forefront.

Table 7. *Content of essential amino acids in proteins from different sources (after Champagnat et al., 1964)*

Protein content of dry substance in %	Wheat flour	Beef	Cow's milk	Fodder yeast	Yeast from petroleum
	13.2	59.4	33.1	44.4	43.6
		Content of amino acids in %			
Leucine	7.0	8.0	11.0	7.6	7.0
Isoleucine	4.2	6.0	7.8	5.5	3.05
Valine	4.1	5.5	7.05	6.0	8.4
Threonine	2.7	5.0	4.7	5.4	9.1
Methionine	1.5	3.2	3.2	0.8	1.2
Lysine	1.9	10.0	8.7	6.8	11.6
Arginine	4.2	7.7	4.2	4.1	8.0
Histidine	2.2	3.3	2.6	1.7	8.1
Phenylalanine	5.5	5.0	5.5	3.9	7.9
Tryptophan	0.8	1.4	1.5	1.6	1.17
Cystine	1.9	1.2	1.0	1.0	0.1

Table 8. *Vitamin content of some foods, as compared with the daily requirement of man (Champagnat et al., 1964)*

Vitamin	Beef mg/kg	Milk mg/kg	Fodder yeast mg/kg d.w.	Yeast from petroleum mg/kg d.w.	Daily requirement of man in mg
B_1, Thiamine	1— 3	0.3— 0.7	2— 20	3— 16	2
B_2, Riboflavine	2	1 — 3	30— 60	75	3
Nicotinic acid	4—100	1 — 5	200—500	180—200	15
Pantothenic acid	7— 21	1 — 4	30—200	150—192	3
B_6, Pyridoxal phosphate	1— 4	1 — 3	40— 50	23	2
B_{12}, Cobalamin				0.11	0.01

At the present state of technology, the biomass produced must be added to animal feeds, because the resulting product – a more or less tasteless white powder – hardly offers any stimulus for direct consumption. Although it has repeatedly been stressed that this is a very valuable food from the point of view of the physiology of nutrition, the resulting biomass must still be considered as crude product which requires refining.

Tables 7 and 8 illustrate the high value of yeast from petroleum-biological processes from the viewpoint of the physiology of nutrition. Among the essential amino acids, one may emphasize the high content of lysine (11.6%), which is almost absent in wheat flour. Thus, there is a physiologically valuable possibility of supplementing protein from higher plants. Moreover, the vitamin content is in some cases very much higher than in other foodstuffs.

5. Outlook

For satisfactory industrial production of single-cell protein from hydrocarbons, it must be possible to manufacture the product in the economically most favourable manner. A compilation of the costs in hydrocarbon fermentation is as follows (Wang, 1968).

Table 9. *Analysis of production costs (Wang, 1968)*

	%
Raw materials (excluding feedstock)	30
Fermentation	15
Separation, drying and packing	15
Staffing and maintenance	15
Depreciation (including interest on capital)	25
	100

The item fermentation covers the expensive devices for controlling the complex reaction system and the high expenditure for oxygen supply. The total costs are, however, determined by the expense of making the product edible (separation, preparation, drying, further processing).

For the present, development has proceeded as far as fodder production. But the ultimate aim is a product which can be used directly as food for man. Currently, the requirements of the health authorities are prohibitive. But research workers are intensively engaged in examining the questions of food hygiene. Since the problems of separation can be solved very easily with methane or methanol, such substrates have in recent years offered a better chance as the basis for single-cell protein production.

Furthermore, the biosynthesis of special compounds on the basis of hydrocarbons has received much attention. As shown in Table 5, above all, research on amino acids and nucleic acids is under way. It may well be that in future, along with the existing petrochemistry, there will also develop a specific petro-biochemistry.

References

Aiba, S., Moritz, V., Someya, J., Haung, K. L.: J. Ferment. Technol. **47**, 203 (1969a).
— Haung, K. L., Moritz, V., Someya, J.: J. Ferment. Technol. **47**, 211 (1969b).
Aida, T., Yamaguchi, K.: Agr. Biol. Chem. **33**, 1244 (1969).
Auliffe,Mc, C.: Nature (London), **200**, 1092 (1963).
Azoulay, E., Chouchoud-Beaumont, P., Senez, J. C.: Ann. Inst. Pasteur **107**, 520 (1964).
— Chouteau, J., Davidovics, G.: Biochim. Biophys. Acta **77**, 554 (1963).
Baker, E. G.: ACS Petroleum Div. Reprints 1 (April 1956).
Bakhuis, E., Bos, P.: Ant. van Leeuwenhock 35, Supplement: Yeast Symposium F 47 (1969).
Buswell, J. A., Jurtshuk, P.: Arch. Mikrobiol. **64**, 215 (1969).
Calderbank, P. H.: In: Biochem. and Biolog. Sci., N. Blakebrough (Ed.), p.102, 1967.
Champagnat, A., Vernet, C., Lainé, B., Filosa, J.: Nature (London) **197**, 13 (1963).
Davis, J. B., Updegraff, D. M.: Bact. Rev. **18**, 215 (1954).
Dostalek, M., Munk, V., Volfova, O., Pecka, K.: Biotechnol. Bioeng. **10**, 33 (1968).
Dunlap, K. R., Perry, J. J.: J. Bacteriol. **94**, 1919 (1967).
Dunn, I. J.: Biotechnol. Bioeng. **10**, 891 (1968).
Einsele, A., Fiechter, A.: Pathol. Microbiol. **34**, 149 (1969).
Erdtsieck, A., Rietema, K.: Ant. van Leeuwenhoek 35, Supplement: Yeast Symposium F 19 (1969).
Erickson,L.E., Humphrey, A. E., Prokop, A.: Biotechnol. Bioeng. **11**, 449 (1969a).
— — Biotechnol. Bioeng. **11**, 467 (1969b).
— — Biotechnol. Bioeng. **11**, 489 (1969c).
Ermakova, I. T., Rozenfeld, S. M., Novakovskaya, N. S., Neklyudova, L. V., Disler, E. N.: Prikl. Biochim. Mikrobiol. **5**, 252 (1969).
Eyk van, J., Bartels, T. J.: J. Bacteriol. **96**, 706 (1968).
Fencl., Z.: In: Symposium on Membrane Transport and Metabolism, p. 296. Czechoslovak Academy of Sciences, Prague 1961.
Fritsche, W.: Zeitschrift Allg. Mikrobiologie **8**, 91 (1968).
Fuhs, G. W.: Arch. Mikrobiol. **39**, 374 (1961).
Glikmans, G.: Rapport IFP Nr. **16**, 019 (1968).
Gradova, N. B., Tretyakova, V. P., Kruchinina, L. K.: In: Proceedings of the 4th Symposium on cont. cult. of microorganisms, p. 569. Academia, Prague 1969.
Hopkins, S. J., Chibnall, A. C.: Biochem. J. **26**, 133 (1932).
Humphrey, A. E.: Biotechnol. Bioeng. **9**, 3 (1967).
— Chem. Can. **20**, 28 (1968).
Iizuka, H., Hirano, A., Shiio, T., Yamamoto, K., Tsuru, S.: Repts. Inst. Appl. Mikrobiol., Tokyo, 252 (1966a).
— Iida, M., Toyoda, S.: Zeitschrift Allg. Mikrobiol. **6**, 335 (1966b).

Ikeda, S., Hirose, Y., Kobayashi, K., Kinoshita, K.: Agr. Biol. Chem. 33, 1042 (1969).
Imada, Y., Takahashi, J., Yamada, K., Uchida, K., Aida, K.: Biotechnol. Bioeng.
    9, 45 (1967).
Johnson, M. J.: Chem. and Ind. Sept. 5, 5 (1964).
Kenna Mc, E. J., Kallio, R. E.: Ann. Rev. Microbiol. 19, 183 (1965).
Klug, M. J., Markovetz, J.: Appl. Microbiol. 15, 690 (1967).
— — Biotechnol. Bioeng. 11, 427 (1969).
Kobayashi, K., Ikeda, S., Irose, Y., Kinoshita, K.: Agr. Biol. Chem. 31, 1448 (1967).
Kyowa Hakko Kogyo KK FR 2000641 P. 5.12.69 (1969).
Linden, A. C. van der, Thijsse, G. J. E.: Adv. in Enzymology 27, 469 (1965).
Ludvik, J., Munk, V., Dostalek, M.: Experientia 24, 1066 (1968).
Mabe, J. A., Brannon, D. R.: Bact. Proc. A 23 (1970).
Makula, R., Finnerty, W. R.: J. Bacteriol. 95, 2102 (1968).
Mimura, A., Kawano, T., Kodaira, R.: J. Ferment. Technol. 47, 229 (1969).
Munk, V., Volfova, O., Dostalek, M., Mostecky, J., Pecka, K.: Folia Micro-
    biologica 14, 334 (1969).
Mycowski 1895: In: Soil and Microbiology. Iizuka, H., Komagata, K. (Ed.), 1961.
Noyes, R.: Food Processing Review No. 3: Protein Food Supplements, Noyes
    Development Corporation, New Yersey, USA (1969).
Nyns, E. J., Auquiere, J., Wiaux, A. L.: Zeitschrift Allg. Mikrobiol. 9, 373 (1969).
— — — Ant. van Leeuwenhoek 34, 441 (1968).
Popova, T. E.: Prikl. Biokhim. Mikrobiol. 4, 103 (1968).
Ratledge, E.: Biotechnol. Bioeng. 10, 511 (1968).
Raymond, R. L., Jamison, V. W., Hudson, J. O.: Appl. Microbiol. 15, 357 (1967).
— — — Appl. Microbiol. 17, 512 (1969).
Ribbons, D. W.: Bact. Proc. 2, A 13 (1969).
Shimizu, S., Tanaka, A., Fukui, S.: J. Ferment. Technol. 47, 551 (1969).
Suomalainen, H.: Suomen Kemistileti A 41, 239 (1968).
Swisher, R. D.: J. Amer. Oil Chem. Soc. 40, 648 (1963).
Tanaka, A., Maki, H., Fukui, S.: J. Ferment. Technol. 45, 1156 (1967a).
— — — J. Ferment. Technol. 45, 1163 (1967b).
— Nagasaki, T., Fukui, S.: J. Ferment. Technol. 46, 477 (1968).
Tokoro, Y'oh, Oshima, K., Tanaka, K., Kinoshita, S.: Amino Acid and Nucleic
    Acid. 19, 115 (1969).
Wagner, F., Kleemann, Th., Zahn, W.: Biotechnol. Bioeng. 11, 393 (1969).
Wang, D. I. C.: In: Single-Cell Protein. Mateles, R. J., Tannenbaum, S. R. (Ed.),
    p. 217. M.I.T. Press, 1968.
Wodzinski, R. S.: Ph. D. thesis, University of Wisconsin (1968).
Yagi, O., Yamada, K.: Agr. Biol. Chem. 33, 1587 (1969).
Yamada, K., Takahashi, J., Kawabata, Y., Okada, T., Onihara, T.: In: Single-Cell
    Protein. Mateles, R. J., Tannenbaum, S. R. (Ed.), p. 192. M.I.T. Press, 1968.
Zobell, C. E.: Bact. Rev. 10, 1 (1946).

A. Fiechter and A. Einsele
Mikrobiologisches Institut
Eidg. Technische Hochschule
CH-8006 Zürich, Weinbergstr. 38